# Moduli of Riemann Surfaces, Real Algebraic Curves, and Their Superanalogs

Translations of
# MATHEMATICAL MONOGRAPHS

Volume 225

# Moduli of Riemann Surfaces, Real Algebraic Curves, and Their Superanalogs

S. M. Natanzon

Translated by
Sergei Lando

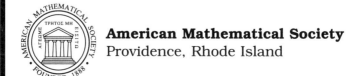

**American Mathematical Society**
Providence, Rhode Island

С. М. Натанзон

## МОДУЛИ РИМАНОВЫХ ПОВЕРХНОСТЕЙ И ВЕЩЕСТВЕННЫХ АЛГЕБРАИЧЕСКИХ КРИВЫХ И ИХ СУПЕРАНАЛОГИ

МЦНМО, МОСКВА, 2003

The work was originally published in Russian by МЦНМО under the title "Модули римановых поверхностей и вещественных алгебраических кривых и их супераналоги" ©2003. The present translation was created under license for the American Mathematical Society and is published by permission.

2000 *Mathematics Subject Classification.* Primary 14H15, 32G15; Secondary 14H40, 14M30, 14P25, 30F35, 30F60, 32C11, 57M12.

For additional information and updates on this book, visit
**www.ams.org/bookpages/mmono-225**

**Library of Congress Cataloging-in-Publication Data**
Natanzon, S. M., 1948–
  [Moduli rimanovykh poverkhnostei i veshchestvennykh algebraicheskikh krivykh i ikh super-analogi English]
  Moduli of Riemann surfaces, real algebraic curves, and their superanalogs / S. M. Natanzon.
    p. cm. — (Translations of mathematical monographs, ISSN 0065-9282 ; v. 225)
  Includes bibliographical references and index.
  ISBN 0-8218-3594-7
  1. Riemann surfaces.   2. Curves, Algebraic.   I. Title.   II. Series.

QA333.N3513   2004
515′.93—dc22                                             2004051990

*To Irina with love and gratitude*

# Contents

# Introduction

This book is devoted to the study of topological properties of the moduli spaces of Riemann surfaces and other moduli spaces close to them: the moduli spaces of algebraic curves and of mappings, as well as superanalogs of all these moduli spaces.

During the entire 20th century, the moduli spaces of Riemann surfaces have attracted permanent attention of mathematicians. In the last two decades, their investigation got a new and powerful incentive due to the discovery of deep connections between the moduli spaces of Riemann surfaces and string theory, which is the modern version of the unified field theory; see [**87**]. The string theory naturally leads to the study of the noncommutative analog of Riemann surfaces, the $N = 1$ super-Riemann surfaces [**27**], [**6**]. The mathematical part of this theory is reduced to the study of "string measure" on the moduli spaces of $N = 1$ super-Riemann surfaces [**7**], [**88**]. The $N = 2$ super-Riemann surfaces arise in the course of further development of the theory [**25**], [**29**].

According to standard definitions, a *real algebraic curve* is a complex algebraic curve (that is, a compact Riemann surface) $P$ equipped with a real structure (this means that an antiholomorphic complex conjugation involution $\tau \colon P \to P$ is given). The category of real algebraic curves is isomorphic to the category of Klein surfaces [**1**], [**66**]. The study of real algebraic curves was initiated by Klein [**40**]. During a long period afterwards, researchers were mainly interested in plane (that is, embedded in either projective or affine plane) algebraic curves. Researchers returned to the systematic study of "general" real algebraic curves only in the 1970s [**24**], [**1**], [**53**], [**54**], [**55**], [**90**]. The method of algebro-geometric integration of equations of mathematical physics discovered in the 1970s by S. P. Novikov and his school led to the appearance of new problems in the theory of real curves and seriously stimulated its further development [**14**], [**21**], [**18**], [**22**], [**65**], [**71**], [**78**]. Another area of applications is the conformal field theory, in particular, string theory [**13**], [**91**], [**13**].

A natural generalization of the moduli spaces are the moduli spaces of mappings of algebraic curves and Riemann surfaces. In the case of constant mappings, these spaces coincide with the moduli spaces of curves. In the case of mappings to the sphere they coincide with the spaces of meromorphic functions studied by Hurwitz [**32**]. It has been understood recently that spaces of holomorphic mappings play the central role in the two-dimensional field theory: presumably semisimple Frobenius manifolds in the sense of Dubrovin [20] can be reduced to these spaces and their generalizations.

In particular, spaces of quantum cohomology of even degree carry natural structures of Frobenius manifolds [41], [89]. Investigation of quantum cohomology of odd degree requires a superanalog of the notion of Frobenius manifold [41], [46]. This fact makes it necessary to study spaces of superholomorphic mappings of super-Riemann surfaces. Another source of superholomorphic mappings is provided by superholomorphic instantons [48], [49], [50].

Real (that is, preserving the real structure) holomorphic mappings of real algebraic curves also appear in many problems in mathematics and mathematical physics. For example, real mappings to the Riemann sphere (real meromorphic functions) play a crucial role in the theory of matrix finite-gap differential operators [97].

In the present book, we study the topology of the moduli spaces of (super-) Riemann surfaces and real algebraic (super) curves, as well as the topology of the space of (super) holomorphic mappings of Riemann surfaces, including those respecting the real structure. We also investigate topological properties of meromorphic tensor fields on real algebraic curves and of the $\theta$-divisor of real algebraic curves that are important for applications.

As a rule, we consider only hyperbolic Riemann surfaces and algebraic curves of genus greater than 1. The case of nonhyperbolic surfaces and curves (spheres with less than three punctures and tori) is simpler, but requires different approaches. The construction of topological invariants of super-Riemann surfaces and algebraic supercurves is based on the study of families of Arf functions. The real-analytic structure of the connected components of the moduli spaces of (super-) Riemann surfaces, algebraic (super) curves and their mappings is studied by means of the theory of (super-) Fuchsian groups.

**The first chapter** is called "Moduli of Riemann surfaces, spaces of Hurwitz type, and their superanalogs." It consists of 15 sections.

The first two sections are devoted to the study, following [38], [52], [55], of the standard generators of Fuchsian groups on the Lobachevsky plane $\Lambda$. Basing on the results of Sections 1 and 2 we turn to the study of the moduli space $M_{g,k,m}$ of (biholomorphic equivalence classes of) Riemann surfaces of genus $g$ with $k$ holes and $m$ punctures.

In Sections 3 and 4, we investigate the spaces $T_{g,k,m}$ of standard generators of Fuchsian groups $\Gamma$ such that $\Lambda/\Gamma \in M_{g,k,m}$. We construct a special parametrization of $T_{g,k,m}$ in a way which is convenient for our purposes. Using this parametrization, we prove the following analog of the classical Frike–Klein theorem [28]: the space $T_{g,k,m}$ is homeomorphic to $\mathbb{R}^{6g+3k+2m-6}$. In Section 5 we study the action on $T_{g,k,m}$ of the group $\mathrm{Mod}_{g,k,m}$ of homotopy classes of autohomeomorphisms of a surface $P \in M_{g,k,m}$ and reproduce the Frike–Klein theorem [28], which states that $M_{g,k,m} = T_{g,k,m}/\mathrm{Mod}_{g,k,m}$ and the action of $\mathrm{Mod}_{g,k,m}$ is discrete.

Section 6 is devoted to holomorphic mappings (morphisms) of one Riemann surface onto another. Two such mappings $f_1 \colon \tilde{P}_1 \to P_1$, $f_2 \colon \tilde{P}_2 \to P_2$ are considered the same if there are biholomorphic mappings $\tilde{\phi} \colon \tilde{P}_1 \to \tilde{P}_2$, $\phi \colon P_1 \to P_2$ such that $\phi f_1 = f_2 \tilde{\phi}$. Two morphisms $(\tilde{P}_1, f_1, P_1)$, $(\tilde{P}_2, f_2, P_2)$ are said to be *topologically equivalent* if there are homeomorphisms $\tilde{\phi} \colon \tilde{P}_1 \to \tilde{P}_2$ and $\phi \colon P_1 \to P_2$ such that $\phi f_1 = f_2 \tilde{\phi}$. The topological equivalence class of a morphism is called its *topological type*. Fix a topological type $t$ and consider the set $H'$ of all morphisms belonging to this topological type. We introduce a natural topology on the set $H'$ and prove that

*The space $H'$ is homeomorphic to $\mathbb{R}^n / \mathrm{Mod}$, where $\mathrm{Mod}$ is a discrete group.*

Think of the Lobachevsky plane $\Lambda$ as of the upper half-plane $\{z \in \mathbb{C} | \mathrm{Im}\, z > 0\}$. Then the automorphism group $\mathrm{Aut}(\Lambda)$ coincides with $\mathrm{PSL}(2, \mathbb{R})$. Denote by

$$J \colon \mathrm{SL}(2, \mathbb{R}) \to \mathrm{PSL}(2, \mathbb{R}) = \mathrm{Aut}(\Lambda)$$

the natural projection. In Section 7 we study, for an arbitrary Riemann surface $P \in M_{g,k,m}$, the subgroups $\Gamma^* \subset \mathrm{SL}(2, \mathbb{R})$ such that $\mathrm{Ker}\, J \cap \Gamma^* = 1$ and $P = \Lambda / J(\Gamma^*)$. We prove that such subgroups are in one-to-one correspondence with Arf functions on $P$, that is, with mappings $\psi \colon H_1(P, \mathbb{Z}_2) \to \mathbb{Z}_2$ (where $\mathbb{Z}_2 = \mathbb{Z}/2\mathbb{Z} = \{0, 1\}$) such that

$$\psi(a + b) = \psi(a) + \omega(b) + (a, b)$$

(here $(.\,,.)$ is the intersection pairing in $H_1(P, \mathbb{Z}_2)$).

In Sections 8 and 9 we study topological invariants of Arf functions and tuples $(\omega_1, \ldots, \omega_n)$ of Arf functions. According to [4], [51], Arf functions $\omega$ on compact surfaces are in one-to-one correspondence with $\theta$-characteristics, and their only topological invariant $\mathrm{Arf}(\omega) \in \mathbb{Z}_2$ coincides with the parity of the corresponding $\theta$-characteristic. On noncompact surfaces the functions $\omega$ possess additional topological invariants; the latter are described in Section 8. A complete system of topological invariants of a pair $(\omega_1, \omega_2)$, of which we make use further, also is derived in this section. In Section 9 we prove that the set of numbers $\{\mathrm{Arf}(\omega_i + \omega_j + \omega_r) \,|\, i, j, r = 1, \ldots, n\}$ forms a complete system of topological invariants of a nondegenerate system of Arf functions $(\omega_1, \ldots, \omega_n)$.

In Section 10 we use the results of the previous sections to describe the topological structure of the space $S_{g,k,m}$ of pairs $(P, e)$, where $P \in M_{g,k,m}$ and $e \colon E \to P$ is a line spinor bundle. We prove that the connected components of $S_{g,k,m}$ are in one-to-one correspondence with the topological types of Arf functions on $P \in M_{g,k,m}$. (For compact surfaces, this statement follows from [4], [51].) Then we prove that

*A connected component $S$ of the space $S_{g,k,m}$ is homeomorphic to the space $T_{g,k,m} / \mathrm{Mod}^S$, where $\mathrm{Mod}^S \subset \mathrm{Mod}_{g,k,m}$.*

The last sections of the chapter are devoted to super-Riemann surfaces. We follow the approach of [6] and [45] according to which super-Riemann surfaces can be represented as superspaces of orbits $P = \Lambda^S/\Gamma$ of a super-Fuchsian group $\Gamma$ acting on the Lobachevsky superplane $\Lambda^S$. An ordinary Riemann surface $P^\sharp$ is associated to any super-Riemann surface $P$.

In Section 11 we show that each $N = 1$ super-Riemann surface $P$ generates an Arf function $\omega_P \colon H_1(P^\sharp, \mathbb{Z}_2) \to \mathbb{Z}_2$. Moreover, for each Arf function $\omega \colon H_1(Q, \mathbb{Z}_2) \to \mathbb{Z}_2$ on a Riemann surface $Q$ there is a super-Riemann surface $P$ such that $P^\sharp = Q$ and $\omega_P = \omega$. We term the topological type $t$ of the Arf function $\omega_P$ the *type of the super-Riemann surface $P$*. Associated to this type is the topological type $\sharp t = (g, k, m)$ of the corresponding Riemann surface $P^\sharp \in M_{g,k,m}$.

In Section 12 we prove the following statement:

*The set of $N = 1$ super-Riemann surfaces of arbitrary topological type $t$, where $\sharp t = (g, k, m)$, can be represented in the form $T^t/\mathrm{Mod}^t$, where the superspace $T^t$ is strongly diffeomorphic to the superspace*

$$\mathbb{R}^{(6g+3k+2m-6\,|\,4g+2k+2m-4)}/\mathbb{Z}_2$$

*and the group $\mathrm{Mod}^t \subset \mathrm{Mod}_{g,k,m}$ acts discretely.*

Sections 13 and 14 are devoted to $N = 2$ super-Riemann surfaces. In Section 13 we associate to an arbitrary $N = 2$ super-Riemann surface $P$ a pair of Arf functions $(\omega_P^1, \omega_P^2)$ and show that for each pair of Arf functions $(\omega^1, \omega^2)$ on a Riemann surface $Q$ there is an $N = 2$ super-Riemann surface $P$ such that $P^\sharp = Q$ and $(\omega_P^1, \omega_P^2) = (\omega^1, \omega^2)$. We call the topological type $t$ of the pair of Arf functions $(\omega_P^1, \omega_P^2)$ *the topological type of the $N = 2$ super-Riemann surface $P$*.

In Section 14 we prove that

*The set of $N = 2$ super-Riemann surfaces of arbitrary topological type $t$ admits a representation of the form $T^t/\mathrm{Mod}^t$, where the superspace $T^t$ is strongly diffeomorphic to the superspace $R^{(a|b)}/(\mathbb{Z}_2)^2$ and the group $\mathrm{Mod}^t \in \mathrm{Mod}_{\sharp t}$ acts discretely.*

We also describe the dependence of the dimensions $(a|b)$ and the group $\mathrm{Mod}^t$ on $t$.

The last Section 15 is devoted to supercoverings, that is, superholomorphic mappings $f \colon \tilde{P} \to P$ of $N = 1$ super-Riemann surfaces that induce holomorphic (unramified) coverings $f^\sharp \colon \tilde{P}^\sharp \to P^\sharp$. Two superholomorphic morphisms $f_1 \colon \tilde{P}_1 \to P_1$ and $f_2 \colon \tilde{P}_2 \to P_2$ are considered the same if there are superholomorphic isomorphisms $\tilde\phi \colon \tilde{P}_1 \to \tilde{P}_2$ and $\phi \colon P_1 \to P_2$ such that $\phi f_1 = f_2 \tilde\phi$. We say that $f_1$ and $f_2$ have coinciding topological type if there are homeomorphisms $\tilde\psi^\sharp \colon \tilde{P}_1^\sharp \to \tilde{P}_2^\sharp$ and $\psi^\sharp \colon P_1^\sharp \to P_2^\sharp$ such that $\psi^\sharp f_1^\sharp = f_2^\sharp \tilde\psi^\sharp$ and $\omega_{P_1} = \omega_{P_2} \psi^\sharp$.

We prove that

*The set of supercoverings of a given topological type $t$ forms a connected supermanifold of the form $T^t/\mathrm{Mod}_t$, where $T^t$ is a superspace strongly diffeomorphic to $\mathbb{R}^{(a|b)}/\mathbb{Z}_2$ and $\mathrm{Mod}_t$ is a discrete group.*

**The second chapter** is called "Moduli of real algebraic curves and their superanalogs. Differentials, spinors, and Jacobians of real curves." It consists of 14 sections.

In Section 1 we describe, following [**93**], the topological invariants $(g, k, \varepsilon)$ of a real algebraic curve $(P, \tau)$. These are the genus $g$ of $P$, the number $k$ of connected components (ovals) of the set $P^\tau = \{p \in P \mid \tau p = p\}$ of real points of $(P, \tau)$, and the number

$$\varepsilon(P, \tau) = \begin{cases} 0 & \text{if } P \setminus P^\tau \text{ is connected } (\textit{nonseparating curve}), \\ 1 & \text{if } P \setminus P^\tau \text{ is connected } (\textit{separating curve}). \end{cases}$$

The only restrictions on the set $(g, k, \varepsilon)$ are $0 \leqslant k \leqslant g$ for $\varepsilon = 0$ and $1 \leqslant k \leqslant g + 1$, $k \equiv g + 1 \pmod 2$ for $\varepsilon = 1$. If $k = g + 1$, then the curve $(P, \tau)$ is called an $M$-$curve$.

In Section 2, using the results of Sections 1–4 of Chapter 1, we prove that

*The set of real algebraic curves of topological type $(g, k, \varepsilon)$ is a connected space of the form $T_{g,k,\varepsilon} / \mathrm{Mod}_{g,k,\varepsilon}$, where $T_{g,k,\varepsilon}$ is homeomorphic to $\mathbb{R}^{3g-3}$, and $\mathrm{Mod}_{g,k,\varepsilon}$ is a discrete group.*

An equivalent statement was first formulated in Earle's paper [**24**]. However, the proof given there is based on the paper [**42**], which contains a mistake. The first correct proof (based on different ideas) was given in [**53**], [**54**], [**55**]. Another independent proof can be found in [**90**].

Section 3 is devoted to Arf functions for real algebraic curves $(P, \tau)$, that is, Arf functions $\omega \colon H_1(P, \mathbb{Z}_2) \to \mathbb{Z}_2$ such that $\omega\tau(a) = \omega(a)$ for all $a \in H_1(P, \mathbb{Z}_2)$. This notion, important in further research, is equivalent to that of a real $\theta$-characteristic in the sense of [**30**]. We construct a set of topological invariants of Arf functions on $(P, \tau)$; the completeness of this set is proved in Section 11. For every such set (called a topological type) the number of the corresponding Arf functions is computed.

Now let $\mathrm{SL}_\pm(2, \mathbb{R}) = \{A \in \mathrm{GL}(2, \mathbb{R}) \mid \det A = \pm 1\}$ and let $\widetilde{\mathrm{Aut}}(\Lambda)$ be the isometry group of the Lobachevsky plane $\Lambda = \{z \in \mathbb{C} \mid \mathrm{Im}\, z > 0\}$. Extend the epimorphism $J \colon \mathrm{SL}(2, \mathbb{R}) \to \mathrm{Aut}(\Lambda)$ constructed in Section 7 of Chapter 1 to an epimorphism $J \colon \mathrm{SL}_\pm(2, \mathbb{R}) \to \widetilde{\mathrm{Aut}}(\Lambda)$ by setting $J \left( \begin{smallmatrix} -1 & 0 \\ 0 & 1 \end{smallmatrix} \right) = -\bar{z}$. In Section 4 we study groups $\tilde{\Gamma}^* \subset \mathrm{SL}_\pm(2, \mathbb{R})$ such that $\mathrm{Ker}\, J \cap \tilde{\Gamma}^* = 1$ and $\Lambda / J(\tilde{\Gamma}^*) = P/\langle \tau \rangle$. We describe their relationships with the Arf functions on $(P, \tau)$. Besides, we prove that the group $\tilde{\Gamma}^*$ defines an orientation on the ovals of $(P, \tau)$.

The results of Sections 3 and 4 allow us to establish, in Section 5, a one-to-one correspondence between real (that is, invariant with respect to the antiholomorphic involution) spinor bundles over a real curve $(P, \tau)$ and nonsingular Arf functions on $(P, \tau)$. Holomorphic sections of a real spinor bundle that are symmetric with respect to the antiholomorphic involution $\tau$ are called *real spinors*. We prove that the real spinors induce an orientation on the ovals of a real curve. Moreover, this orientation as well as the parity of the number of zeroes of a spinor on an oval is determined by the values of the corresponding Arf function on some explicitly described cycles. This fact allows one to construct real spinors with prescribed properties.

Section 6 is devoted to real differentials on real curves $(P, \tau)$, that is, holomorphic differentials on $P$ symmetric with respect to $\tau$. Let us orient the ovals $P^\tau$ of a real algebraic curve in such a way that if $\varepsilon(P, \tau) = 1$, then the orientation of $P^\tau$ is induced by the orientation of one of the connected components of the complement $P \setminus P^\tau$. A symmetric (i.e., such that $\tau u(p) = \overline{u(p)}$) local chart $z \colon U \to \mathbb{C}$ takes the real points $P^\tau \cap U$ of an oval $c \subset P^\tau$ to the real line $\mathbb{R} \subset \mathbb{C}$. Choose a local chart in such a way that the orientation of the oval $c$ produces on the set $z(P^\tau \cap U)$ the orientation coinciding with the standard orientation of $\mathbb{R}$. In such a local chart a real differential takes real values on the oval, and the sign of the values is independent of the choice of the local chart. The results of Section 5 allow us to prove that on nonseparating curves the real holomorphic differentials realize all combinations of signs on the ovals. We also study restrictions that arise on separating curves. In particular, we prove that

*For any real differential on an M-curve, there is an oval where it is totally positive and an oval where it is totally negative.*

In Section 7 we construct, on an arbitrary separating real algebraic curve, the real spinors $\sin_n^\lambda(x), \cos_n^\lambda(x)$ of weight $\lambda \in \mathbb{Z} \cup (\mathbb{Z} + 1/2)$, which generalize classical trigonometric functions. In their construction, the "symmetrization" of tensor fields used by Krichever and Novikov [**43**] in their construction of the conformal field theory on complex algebraic curves is applied. For the functions thus constructed we prove analogs of summation theorems for trigonometric functions, of Fourier theorems about series expansions for real periodic functions (real tensors, in our case), and the following analog of the Sturm–Hurwitz theorem [**33**]:

*Let $(P, \tau)$ be a real algebraic curve of type $(g, k, 1)$. Suppose that either $\lambda \neq 0, 1$ or $n > \frac{g}{2}$. Then the tensor $F = \sum_{i=n}^{\infty}(a_i \cos_i^\lambda + b_i \sin_i^\lambda)$ $(a_i, b_i \in \mathbb{R})$ has at least $2n - g$ zeros on the ovals $P^\tau$.*

In Section 8 we make use of the properties of real holomorphic differentials established in Section 6 in order to describe real and imaginary tori of the Jacobian of an arbitrary real curve. We describe these tori explicitly and indicate those of them that intersect the $\theta$-divisor. This was done first in [**26**] for separating curves and in [**21**] for nonseparating ones. These results were used afterwards in the description of solutions of the Kadomtsev–Petviashvili equations playing an important role in physics [**22**].

In Section 9 we use the results of Sections 6 and 8 to obtain a description of the real tori of the Prymians of real curves $(P, \tau)$ possessing a holomorphic involution $\alpha \colon P \to P$, $\alpha \tau = \tau \alpha$, with exactly two fixed points. In this case $(P/\langle \alpha \rangle, \tau/\langle \alpha \rangle)$ is a real curve of some type $(\tilde{g}, \tilde{k}, \tilde{\varepsilon})$. In particular, we prove that

*For $\tilde{\varepsilon} = 0$ all real tori of the Prymian intersect the $\theta$-divisor. For $\tilde{\varepsilon} = 1$ at most one real torus intersects the $\theta$-divisor, and for $\tilde{k} = \tilde{g} + 1$ such a torus always exists.*

These results are used in [65] to describe nonsingular potentials of real two-dimensional Schrödinger operators of Veselov–Novikov type. The parametrization of the Frike–Klein–Teichmüller space constructed in Sections 1–4 of Chapter 1 was used by Bobenko [9], [10] in his construction of a "Schottky type" parametrization of spaces of real algebraic curves. In Section 10 we reproduce the Bobenko constructions. Their analog for curves with involutions was used in the construction of two-dimensional Schrödinger operators in [65].

We return to spinors in Section 11, where we prove that

*The connected components of the moduli spaces of the real spinor bundles of rank 1 are in one-to-one correspondence with the topological types of nondegenerate Arf functions (defined in Section 3). Each connected component $S$ is diffeomorphic to $\mathbb{R}^{3g-3} / \mathrm{Mod}_S$, where $g$ is the genus of the curves $P$, and $\mathrm{Mod}_S$ is a discrete group.*

The last three sections of Chapter 2 are devoted to real algebraic supercurves. We define the latter as pairs $(P, \tau)$, where $P$ is a super-Riemann surface with a compact number part $P^\sharp$, and $\tau \colon P \to P$ is a superantiholomorphic involution.

In Section 12 we associate to a real $N = 1$ supercurve $(P, \tau)$ a real algebraic curve $(P^\sharp, \tau^\sharp)$ and an Arf function $\omega = \omega(P, \tau)$ on $(P^\sharp, \tau^\sharp)$. The topological type of the Arf function $\omega$ is called the *topological type of the supercurve* $(P, \tau)$. We prove that

*The connected components of the moduli space of real $N = 1$ supercurves are in one-to-one correspondence with the topological types of nonsingular real Arf functions. Each connected component $S$ has the form $T / \mathrm{Mod}_S$, where $T$ is strongly diffeomorphic to the superspace $\mathbb{R}^{(3g-3|2g-2)} / \mathbb{Z}_2$, $g$ is the genus of the corresponding supercurves, and $\mathrm{Mod}_S$ is a discrete group.*

In Section 13 we describe a system of invariants of real $N = 2$ supercurves, which we term the *topological type of real algebraic $N = 2$ supercurves*. We prove in Section 14 that

*The topological types are in one-to-one correspondence with the connected components of the moduli spaces of real $N = 2$ supercurves. Each connected component $S$ has the form $T / \mathrm{Mod}_S$, where $T$ is strongly diffeomorphic to the superspace $\mathbb{R}^{(a|b)} / (\mathbb{Z}_2)^2$, and $\mathrm{Mod}_S$ acts discretely.*

We also describe the dependence of the dimension $(a|b)$ and the group $\mathrm{Mod}_S$ on the type of $S$.

**The concluding chapter** is called "Spaces of meromorphic functions on complex and real algebraic curves." It consists of five sections.

A meromorphic function on a complex algebraic curve $P$ can be regarded as a holomorphic mapping $f \colon P \to S$ to the Riemann sphere $S = \mathbb{C} \cup \{\infty\}$. Two meromorphic functions $(P_1, f_1)$ and $(P_2, f_2)$ are considered the same if there is a biholomorphic mapping $\phi \colon P_1 \to P_2$ such that $f_1 = f_2\phi$. A point $p \in P$ such that $df(p) = 0$ is called a *critical point* of $f$, and its image $f(p) \in S$ is a *critical value* of $f$. The critical values $s \in S$ whose preimage $f^{-1}(s)$ consists of $\deg f - 1$ points are said to be *simple*. The set $H^{\mathbb{C}}_{g,n}$ of all meromorphic functions of genus $g$ and degree $n$ possesses a natural topology. The subset $H^0_{g,n}$ consisting of functions with simple poles and simple critical values is called the *Hurwitz space*. It is an open subset in $H^{\mathbb{C}}_{g,n}$ whose closure coincides with the whole space. According to the classical Hurwitz theorem [32] the space $H^0_{g,n}$ is connected.

The first three sections of Chapter 3 are devoted to the study of generalized Hurwitz spaces $H_g(r_1, \dots, r_k) \subset H^{\mathbb{C}}_{g,n}$ consisting of meromorphic functions with simple critical values, and with polar divisors of the form $r_1 p_1 + \cdots + r_k p_k$. The group $\mathrm{Aut}(\mathbb{C})$ acts on $H_g(r_1, \dots, r_k)$ naturally by shifts $f \mapsto Af + B$ ($A \in \mathbb{C} \setminus 0$, $B \in \mathbb{C}$). According to [19], [20], each space $H_g(r_1, \dots, r_k)$ possesses a natural structure of a Frobenius manifold, and their natural generalizations describe all such manifolds [74].

In the first two sections we prove that the functions in $H_g(r_1, \dots, r_k)$ are pairwise topologically equivalent (in the sense of Section 6 of Chapter 1). Using this fact and the results of Section 6 of Chapter 1, we prove that

*The space $H_g(r_1, \dots, r_k)$ is connected; the space $H_g(r_1, \dots, r_k)/\mathrm{Aut}(\mathbb{C})$ is homeomorphic to $T/\mathrm{Mod}$, where $T \cong \mathbb{R}^{4(g+n-2)-2\sum_{j=1}^k (r_j - 1)}$ and $\mathrm{Mod}$ is a discrete group.*

For the case of Laurent polynomials (rational functions of genus 0 with the polar divisor consisting of two points) this result was obtained earlier by V. I. Arnold by means of deep methods from singularity theory.

Sections 4 and 5 are devoted to the description of the connected components of spaces of real meromorphic functions. A *real meromorphic function* is a triple $(P, \tau, f)$, where $(P, \tau)$ is a real algebraic curve and $(P, f)$ is a meromorphic function such that $f(\tau p) = \overline{f(p)}$. Two functions $(P_1, \tau_1, f_1)$ and $(P_2, \tau_2, f_2)$ are considered to be the same if there is a biholomorphic mapping $\phi \colon P_1 \to P_2$ such that $f_1 = f_2\phi$ and $\phi\tau_1 = \tau_2\phi$.

This definition of a real meromorphic function axiomatizes the properties of a real algebraic function, that is, an algebraic function $f = f(z)$ given by an equation

$$a_0(z)f^n + a_1(z)f^{n-1} + \cdots + a_n(z) = 0,$$

whose coefficients $a_i(z)$ are polynomials with real coefficients. The involution $(f, z) \to (\bar{f}, \bar{z})$ determines an antiholomorphic involution $\tau \colon P \to P$ of the Riemann surface $P$ of $f(z)$, and $f(\tau p) = \overline{f(p)}$. It is easy to show that any real meromorphic function can be constructed in this way.

If $(P, \tau, f)$ is a real meromorphic function, $(P, f) \in H^{\mathbb{R}}_{g,n}$, and $c_1, \dots, c_k$ are the ovals of $(P, \tau)$, then $f(c_j) \subset S_0 = \mathbb{R} \cup \infty$. As a result, we are able to define the degree $i_j$ of $(P, \tau, f)$ on the oval $c_j$. For $(P, \tau) \in M_{g,k,0}$, we define the *topological type* of $(P, \tau, f)$ as the tuple $(g, n, 0 | i_1, \dots, i_k)$. In the case of $(P, \tau) \in M_{g,k,1}$ some new topological invariants arise (in particular, a sign can be assigned to the degree of a function on an oval), which form a tuple $(g, n, 1 | i_1, \ldots, i_k | \chi)$ also called the *topological type of* $(P, \tau, f)$. We describe all possible values of the topological type and prove that

*The set of real meromorphic functions* $(P, \tau, f)$ *of a given type is a connected topological space of dimension* $2(g + n - 1)$, *where* $g$ *and* $n$ *are the genus and the degree of the functions, respectively.*

## Acknowledgments

I am grateful to V. I. Arnold, E. B. Vinberg, S. P. Novikov, and A. S. Schwarz, who introduced me to the problems described in this book.

The work on this book has been supported by the research grants RFBR-01-01-00739, RFBR-02-01-14098, NWO 047.008.005 and INTAS-00-0259.

# CHAPTER 1

# Moduli of Riemann Surfaces, Hurwitz Type Spaces and Their Superanalogs

## 1. Fuchsian groups and their sequential generators

Recall that the upper half-plane

$$\Lambda = \{z \in \mathbb{C} \mid \operatorname{Im} z > 0\}$$

carries the Lobachevsky metric of positive constant curvature, $ds = \frac{|dz|}{\operatorname{Im} z}$. The geodesics of this metric are the semicircles and lines orthogonal to $\mathbb{R}$. The group of holomorphic automorphisms $\operatorname{Aut}(\Lambda) \cong \operatorname{PSL}(2, \mathbb{R})$ coincides with the group of orientation preserving isometries of the Lobachevsky metric.

The set $\operatorname{Aut}(\Lambda) \setminus \{\mathrm{id}\}$ decomposes into the subsets $\operatorname{Aut}_0(\Lambda)$ of elliptic, $\operatorname{Aut}_1(\Lambda)$ of parabolic, and $\operatorname{Aut}_2(\Lambda)$ of hyperbolic automorphisms, having, respectively, 0, 1, or 2 fixed points on the absolute $\partial\Lambda = \mathbb{R} \cup \{\infty\}$.

A *parabolic automorphism* with a finite fixed point $a \in \mathbb{R}$ has the form

$$C(z) = \frac{(1 - a\gamma)z + a^2\gamma}{-\gamma z + (1 + a\gamma)};$$

the value $\gamma$ in this formula is called the *shift parameter*. An automorphism $C$ is said to be *positive* if $\gamma > 0$. In this case $C(r) > r$ for $r \in \mathbb{R} \setminus \{a\}$. It is convenient to visualize the action of an automorphism $C$ on the upper half-plane as shown in Fig. 1.1.1, where $\ell(C) = a$. A *hyperbolic automorphism*

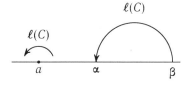

FIGURE 1.1.1

with a finite fixed attracting point $\alpha$ and a finite fixed repelling point $\beta$ has the form

$$C(z) = \frac{(\lambda\alpha - \beta)z + (1 - \lambda)\alpha\beta}{(\lambda - 1)z + (\alpha - \lambda\beta)},$$

where the value $\lambda > 1$ is called the *shift parameter*. An automorphism $C$ is said to be *positive* if $\alpha < \beta$. In this case $C(r) > r$ for $r \in \mathbb{R} \setminus [\alpha, \beta]$ and

$C(r) < r$ for $r \in (\alpha, \beta)$. It is convenient to visualize the action of an auto-morphism $C$ on the upper half-plane as shown in Fig. 1.1.1, where $\ell(C) \subset \Lambda$ is the unique geodesic of the Lobachevsky metric invariant with respect to $C$.

For two automorphisms $C_1, C_2$ with finite fixed points on $\mathbb{R}$, let us agree that $C_1 < C_2$ if each fixed point of $C_1$ is smaller than each fixed point of $C_2$. A set $\{C_1, C_2, C_3\} \subset \mathrm{Aut}_1(\Lambda) \cup \mathrm{Aut}_2(\Lambda)$ is said to be *ordered* if $C_1 \cdot C_2 \cdot C_3 = 1$ and there exists an element $D \in \mathrm{Aut}(\Lambda)$ such that the automorphisms $\tilde{C}_i \doteq DC_iD^{-1}$ ($i = 1, 2, 3$) have finite fixed points, are positive, and $\tilde{C}_1 < \tilde{C}_2 < \tilde{C}_3$. We say that an ordered triple $\{C_1, C_2, C_3\}$ is *sequential* if $\{C_1C_2C_1^{-1}, C_1, C_3\}$ also is an ordered triple. We call a set $\{C_1, \ldots, C_n\}$ *sequential of type* $(0, k, m)$ if for any $i = 1, \ldots, n-1$ the triples $\{C_1 \cdots C_{i-1}, C_i, C_{i+1} \cdots C_n\}$ are sequential, and $n = k + m$, $C_i \in \mathrm{Aut}_2(\Lambda)$ for $i \leqslant k$ and $C_i \in \mathrm{Aut}_1(\Lambda)$ for $i > k$.

A geometric argument looks more convenient not on the upper half-plane $\Lambda$ but on the interior of the unit disk $H = \{z \in \mathbb{C} \mid |z| < 1\}$. The group of its holomorphic automorphisms $\mathrm{Aut}(H)$ also decomposes into el-liptic ($\mathrm{Aut}_0(H)$), parabolic ($\mathrm{Aut}_1(H)$), and hyperbolic ($\mathrm{Aut}_2(H)$) automor-phisms having, respectively, 0, 1 or 2 fixed points on the absolute $\varphi H = \{z \in \mathbb{C} \mid |z| = 1\}$. A biholomorphic isomorphism $\phi \colon \Lambda \to H$ takes the metric on $\Lambda$ to a metric on $H$, which is invariant under $\mathrm{Aut}(H)$; the geodesics of this metric are the circle arcs orthogonal to $\varphi H$. Here

$$\mathrm{Aut}_i(H) = \phi\, \mathrm{Aut}_i(\Lambda)\phi^{-1}.$$

For $C \in \mathrm{Aut}_1(H) \cup \mathrm{Aut}_2(H)$, let $\ell(C) = \phi(\ell(\phi^{-1}C\phi))$. A set $\{C_1, \ldots, C_n\} \in \mathrm{Aut}(H)$ is said to be *sequential* if $\{\phi^{-1}C_1\phi, \ldots, \phi^{-1}C_n\phi\} \in \mathrm{Aut}(\Lambda)$ is a sequential set. According to our agreements, a sequential set looks like the

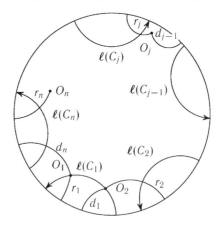

FIGURE 1.1.2

one shown in Fig. 1.1.2, where

$$\ell = C_jC_{j+1} \cdots C_{n-1}C_n(\ell(C_1))$$

lies between $\ell(C_{j-1})$ and $\ell(C_j)$.

Recall that a discrete subgroup of $\text{Aut}(H)$ or $\text{Aut}(\Lambda)$ is called a Fuchsian group. In what follows we shall consider only Fuchsian groups without elliptic elements. Hence, a Fuchsian group is a discrete subgroup of $\text{Aut}(H)$ or $\text{Aut}(\Lambda)$ having no elliptic elements.

LEMMA 1.1. *A sequential set* $V = \{C_1, \ldots, C_n\} \subset \text{Aut}(H)$ *of type* $(0, k, m)$ *generates a Fuchsian group* $\Gamma$. *The surface* $H/\Gamma$ *is a sphere with* $k$ *holes and* $m$ *punctures.*

Proof. Suppose first that $k > 0$. Consider a point $O_1 \subset \ell(C_1)$ and set $O_i = C_i C_{i+1} \cdots C_n(O_1)$ (see Fig. 1.1.2). Let $r_i$ be a geodesic ray starting at $O_i$ and ending at a point of the absolute and intersecting $\ell(C_i)$. Then $d_i = C_i^{-1} r_i$ is a geodesic ray ending at $O_{i+1}$. The rays $\{r_i, d_i \ (i = 1, \ldots, n)\}$ bound a noncompact domain $M$. Our definition guarantees that the sequence of polygons

$$M, \quad C_1 M, \quad C_1 C_2 M, \quad \ldots, \quad C_1 \cdots C_{n-1} M$$

realizes a simple circuit around the point $O_1$. By [95], this implies that $M$ is a fundamental domain for the group $\Gamma$ generated by $\{C_1, \ldots, C_n\}$. The automorphism $C_i$ identifies $r_i$ and $d_i$ thus forming a hole on the surface $H/\Gamma$ if $i \leqslant k$ or a puncture if $i > k$. This construction works also for $k = 0$ if we take for $O_1$ a point sufficiently close to the fixed point of the parabolic automorphism $C_1$.                                                                 □

A *sequential set of type* $(g, k, m)$ is a set

$$\{A_1, B_1, \ldots, A_g, B_g, C_1, \ldots, C_n\}$$

such that $A_i, B_i$ $(i = 1, \ldots, g)$ are hyperbolic automorphisms and

$$\{A_1, B_1 A_1^{-1} B_1^{-1}, \ldots, A_g, B_g A_g^{-1} B_g^{-1}, C_1, \ldots, C_n\}$$

is a sequential set of type $(0, 2g + k, m)$.

A surface of genus $g$ with $k$ holes and $m$ punctures will be called a *surface of type* $(g, k, m)$. We say that a *Fuchsian group* $\Gamma \subset \text{Aut}(H)$ (respectively, $\Gamma \subset \text{Aut}(\Lambda)$) *is of type* $(g, k, m)$ if $H/\Gamma$ (respectively, $\Lambda/\Gamma$) is a surface of type $(g, k, m)$.

THEOREM 1.1 ([**38**], [**52**]). *A sequential set of type* $(g, k, m)$ *generates a group* $\Gamma$ *of type* $(g, k, m)$.

Proof. Let $\{A_1, B_1, \ldots, A_g, B_g, C_{g+1}, \ldots, C_n\} \in \text{Aut}(H)$ be a sequential set of type $(g, k, m)$. For $g = 0$, the statement follows from Lemma 1.1. Now suppose $g > 0$. Set $C_i = [A_i B_i]$ $(i = 1, \ldots, g)$. Our definitions guarantee the location of geodesics $\ell(A_i)$, $\ell(B_i)$, $\ell(C_i)$ shown in Fig. 1.1.3. Let $O_1 \in \ell(C_1)$ and let $M$ be the polygon constructed in the proof of Lemma 1.1.

For $i \leqslant g$, replace the rays $r_i$, $d_i$ by the geodesic segments with the vertices

$$O_i, A_i B_i^{-1} A_i^{-1} O_i, B_i^{-1} A_i^{-1} O_i, A_i^{-1} O_i, B_i A_i B_i^{-1} A_i^{-1} O_i = O_{i+1}.$$

As a result, we obtain a new polygon $\tilde{M}$ (see Fig 1.1.3). Our definitions guarantee that the polygons

$$\tilde{M}, \quad A_1\tilde{M}, \quad A_1B_1\tilde{M}, \quad A_1B_1A_1^{-1}\tilde{M}, \quad C_1\tilde{M}, \quad C_1A_2\tilde{M}, \quad C_1A_2B_2\tilde{M},$$
$$C_1A_2B_2A_2^{-1}\tilde{M}, \quad C_1C_2\tilde{M}, \quad \ldots, \quad C_1C_2\cdots C_{n-1}M$$

realize a circuit around $O_1$ and hence (see [**95**]) the set $\{A_1, \ldots, C_n\}$ generates a Fuchsian group $\Gamma$. It is easy to see that each pair $(A_i, B_i)$ generates a handle of $H/\Gamma$. Therefore, $H/\Gamma$ is a surface of type $(g, k, m)$.    □

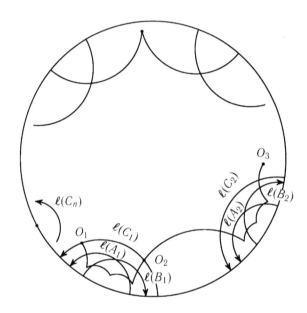

FIGURE 1.1.3

## 2. The geometry of Fuchsian groups

Let $P$ be a surface of type $(g, k, m)$. A system of generators

$$v = \{a_i, b_i \ (i = 1, \ldots, g), c_i \ (i = g+1, \ldots, n)\}$$

of $\pi_1(P, p)$ is said to be *standard* if $v$ generates $\pi_1(P, p)$ with the set of defining relations

$$\prod_{i=1}^{g}[a_i, b_i] \prod_{i=g+1}^{n} c_i = 1$$

and can be represented by a set of simple contours

$$\tilde{v} = \{\tilde{a}_i, \tilde{b}_i \ (i = 1, \ldots, g), \tilde{c}_i \ (i = g+1, \ldots, n)\}$$

possessing the following properties:

(1) the contour $\tilde{c}_i$ is homologous to 0 and encloses a single hole in $P$ for $i \leqslant g + k$ or a single puncture for $i > g + k$;

(2) $\tilde{a}_i \cap \tilde{b}_j = \tilde{a}_i \cap \tilde{c}_j = \tilde{b}_i \cap \tilde{c}_j = \tilde{c}_i \cap \tilde{c}_j = p$;

(3) in a neighborhood of $p$, the mutual positions of the contours $\tilde{v}$ are as shown in Fig. 1.2.1.

In this case the set of contours $\tilde{v}$ is situated on $P$ as shown in Fig. 1.2.2.

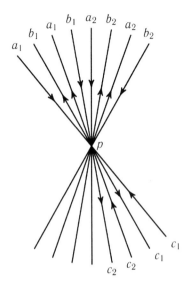

FIGURE 1.2.1

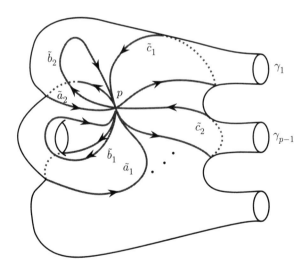

FIGURE 1.2.2

Now let $\Gamma \subset \operatorname{Aut}(H)$ be a Fuchsian group, let $P = H/\Gamma$, let $\Phi \colon H \to P$ be the natural projection, and let $q \in \Phi^{-1}(p)$. We associate to an automorphism $C \in \Gamma$ an oriented geodesic segment $\ell_q(C) \subset H$ starting at $q$ and ending at $C(q)$. The correspondence $C \mapsto \Phi(\ell_q(C))$ generates an isomorphism $\Phi_q \colon \Gamma \to \pi_1(P, p)$.

LEMMA 2.1. *Let $V = \{A_i, B_i \ (i = 1, \ldots, g), C_i \ (i = g+1, \ldots, n)\}$ be a sequential set of type $(g, k, m)$, let $\Gamma$ be the Fuchsian group generated by this set, and let $P = H/\Gamma$. Then $v_q = \Phi_q(V)$ is a standard system of generators of $\pi_1(P, p)$.*

Proof. Consider the fundamental domain $M$ constructed in the course of the proof of Lemma 1.1 and Theorem 1.1 (see Fig. 1.1.3). For $i > g$, let us connect the points $O_i$ and $O_{i+1}$ by pairwise disjoint segments $c_i \subset M$ (here $O_{n+1} = O_1$). Consider the geodesic segments

$$a_i = [O_i, A_i B_i^{-1} A_i^{-1} O_i], \quad b_i = [A_i B_i^{-1} A_i^{-1} O_i, B_i^{-1} A_i^{-1} O_i]$$

on $H$. The natural projection $\Phi \colon H \to P$ produces a standard system of generators

$$V_{O_1} = \{\Phi(a_i), \Phi(b_i) \ (i = 1, \ldots, g), \Phi(c_i) \ (i = g+1, \ldots, n)\} \in \pi_1(P, \Phi(O_1)).$$

The continuous motion of the point $O_1$ to $q$ takes $V_{O_1}$ to a standard system of generators $V_q$. $\square$

The main goal of the present section is to prove the inversion of this lemma.

THEOREM 2.1 ([**52**], [**36**], [**37**], [**38**]). *Let $\Gamma \subset \mathrm{Aut}(H)$ be a Fuchsian group of type $(g, k, m)$, let $P = H/\Gamma$, $\Phi \colon H \to P$ be the natural projection, and let*

$$v = \{a_i, b_i \ (i = 1, \ldots, g), c_i \ (i = g+1, \ldots, n)\}$$

*be the standard system of generators of $\pi_1(P, \Phi(q))$. Then $V = \Phi_q^{-1}(v)$ is a sequential set of type $(g, k, m)$.*

The proof, based on paper [**52**], will require several auxiliary definitions and statements.

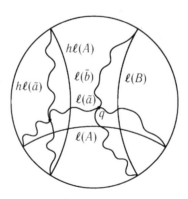

FIGURE 1.2.3

Let $\tilde{a}$ be a contour representing an element $a \in \pi_1(P, \Phi(q))$. We go along $\tilde{a}$ starting at $\Phi(q)$ and lift this path to $H$ starting at $q$. After making infinitely many circuits in both directions we obtain a curve $\ell(\tilde{a}) \subset M$ with ends at the fixed points of the automorphism $A = \Phi_q^{-1}(a)$ (see Fig. 1.2.3).

LEMMA 2.2. *If $\tilde{a}$ has no self-intersections, then $h\ell(A)$ and $\ell(A)$ do not intersect each other for any $h \in \Gamma$.*

Proof. Suppose $h\ell(A)$ and $\ell(A)$ intersect each other for some $h \in \Gamma$. Then $h\ell(\tilde{a})$ and $\ell(\tilde{a})$ also intersect each other (see Fig. 1.2.3). $\qquad\square$

LEMMA 2.3. *Suppose contours $\tilde{a}$, $\tilde{b}$ represent elements $a, b \in \pi_1(P, \Phi(q))$. Suppose there is a small deformation of $\tilde{a}$ and $\tilde{b}$ taking them to disjoint contours. Then*
$$\ell(\Phi_q^{-1}(a)) \cap \ell(\Phi_q^{-1}(b)) = \varnothing.$$

Proof. If
$$\ell(\Phi_q^{-1}(a)) \cap \ell(\Phi_q^{-1}(b)) \neq \varnothing,$$
then $\ell(\tilde{a})$ and $\ell(\tilde{b})$ intersect each other in such a way that their intersection cannot be eliminated by a small deformation (see Fig. 1.2.3). $\qquad\square$

LEMMA 2.4. *Suppose contours $\tilde{c}_1$, $\tilde{c}_2$, $\tilde{c}_3$ have no self-intersections and represent elements $c_1, c_2, c_3 \in \pi_1(P, \Phi(q))$ such that $c_1 \cdot c_2 \cdot c_3 = 1$. Suppose there is a small deformation of the contours $\tilde{c}_1$, $\tilde{c}_2$, and $\tilde{c}_3$ taking them to pairwise disjoint contours. Then either the set*
$$\{\Phi_q^{-1}(c_1), \Phi_q^{-1}(c_2), \Phi_q^{-1}(c_3)\}$$
*or the set*
$$\{\Phi_q^{-1}(c_3^{-1}), \Phi_q^{-1}(c_2^{-1}), \Phi_q^{-1}(c_1^{-1})\}$$
*is sequential.*

Proof. Set $C_i = \Phi_q^{-1}(c_i)$. By Lemma 2.3, $\ell(C_1) \cap \ell(C_2) = \varnothing$. Suppose that the mutual position of $\ell(C_1)$ and $\ell(C_2)$ is as shown in Fig. 1.2.4.

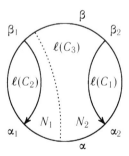

FIGURE 1.2.4

Consider the arcs $\alpha = (\alpha_1, \alpha_2)$ and $\beta = (\beta_1, \beta_2)$ on $\partial H$. Then $C_1\alpha \subset \alpha$, $C_2\alpha \subset \alpha$, whence $C_3^{-1}\alpha = C_1C_2\alpha \subset \alpha$. Similarly, $C_3^{-1}\beta \subset \beta$. Hence $\ell(C_3)$ goes as shown in Fig. 1.2.4 by the dotted line. Then, by Lemma 2.2, $C_1\ell(C_3) \subset N_1$ and, therefore, $C_1\beta_2 \in N_1$. This is impossible because
$$C_1\beta_2 = C_3^{-1}C_2^{-1}\beta_2 = C_3^{-1}\beta \subset N_2.$$

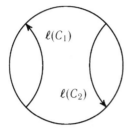

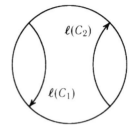

FIGURE 1.2.5                              FIGURE 1.2.6

Thus, $\ell(C_1)$ and $\ell(C_2)$ look like shown either in Fig. 1.2.5 or in Fig. 1.2.6.

A similar statement is valid also for the pairs $(C_2, C_3)$ and $(C_3, C_1)$. Then it follows that either the set $\{C_1, C_2, C_3\}$ or the set $\{C_3^{-1}, C_2^{-1}, C_1^{-1}\}$ is sequential in the sense of Section 1.

Similarly, either the set $\{C_1 C_2 C_1^{-1}, C_1, C_3\}$ or $\{C_3^{-1}, C_1^{-1}, C_1 C_2^{-1} C_1^{-1}\}$ is sequential. Only one of the sets $\{C_1, C_2, C_3\}$ and $\{C_3^{-1}, C_1^{-1}, C_1 C_2^{-1} C_1^{-1}\}$ can be sequential. Hence either the sets $\{C_1, C_2, C_3\}$, $\{C_1 C_2 C_1^{-1}, C_1, C_3\}$ or the sets $\{C_3^{-1}, C_2^{-1}, C_1^{-1}\}$, $\{C_3^{-1}, C_1^{-1}, C_1 C_2^{-1} C_1^{-1}\}$ are sequential.     □

Proof of Theorem 2.1. Set $A_i = \Phi_q^{-1}(a_i)$, $B_i = \Phi_q^{-1}(b_i)$, $C_i = \Phi_q^{-1}(c_i)$. Consider the sets

$$\{x_1, \ldots, x_{g+n}\} = \{a_1, b_1 a_1^{-1} b_1^{-1}, a_2, \ldots, b_g a_g^{-1} b_g^{-1}, c_{g+1}, \ldots, c_n\}$$

and

$$\{X_1, \ldots, X_{g+n}\} = \{A_1, B_1 A_1^{-1} B_1^{-1}, A_2, \ldots, B_g A_g^{-1} B_g^{-1}, C_{g+1}, \ldots, C_n\}.$$

Applying Lemma 2.4 to the set

$$\{x_1 \cdots x_{\ell-1}, x_\ell, x_{\ell+1} \cdots x_{g+n}\},$$

we conclude that either all the sets of the form

$$\{X_1 \cdots X_{\ell-1}, X_\ell, X_{\ell+1} \cdots X_{g+n}\}$$

or all the sets of the form

$$\{X_{g+n}^{-1} \cdots X_{\ell+1}^{-1}, X_\ell^{-1}, X_{\ell-1}^{-1} \cdots X_1^{-1}\}$$

are sequential. This means that either $\{X_1, \ldots, X_{g+n}\}$ or $\{X_{g+n}^{-1}, \ldots, X_1^{-1}\}$ is a sequential set. In the last case, by Lemma 2.1, the contours representing $a_i$, $b_i$, $c_i$, are situated not as in Fig. 1.2.1, and hence it is the set $\{X_1, \ldots, X_{g+n}\}$ that is sequential of type $(0, 2g + k, m)$. Therefore, the set

$$\{A_i, B_i \ (i = 1, \ldots, g), C_j \ (j = g + 1, \ldots, n)\}$$

is sequential of type $(g, k, m)$.     □

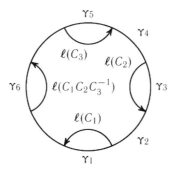

FIGURE 1.3.1

## 3. Free Fuchsian groups of rank $2$

By Theorems 1.1 and 2.1, each sequential set of type $(g, k, m)$ generates a Fuchsian group of type $(g, k, m)$ and each Fuchsian group is generated by such a set.

In this section we give all (up to conjugacy in $\mathrm{Aut}(\Lambda)$) sequential sets of types $(0, 3, 0)$, $(0, 2, 1)$, $(0, 1, 2)$, and $(1, 1, 0)$, thus obtaining a complete description of all Fuchsian groups of rank 2.

LEMMA 3.1. *Each ordered set $\{C_1, C_2, C_3\}$ is sequential.*

Proof. The fixed points of the shifts $C_i$ split the absolute into six arcs $\gamma_i$ as shown in Fig. 1.3.1.

Let $\alpha$ and $\beta$ be the fixed attracting and fixed repelling points, respectively, of the shift $C_1 C_2 C_1^{-1}$. Then

$$\ell(C_1 C_2 C_1^{-1}) = C_1 \ell(C_2) \quad \text{in} \quad \beta \in \gamma_4 \cup \gamma_5 \cup \gamma_6.$$

On the other hand,

$$C_1 \ell(C_2) = C_3^{-1} C_2^{-1} \ell(C_2) = C_3^{-1} \ell(C_2) \quad \text{and} \quad \beta \in \gamma_3 \cup \gamma_2 \cup \gamma_1 \cup \gamma_6.$$

Hence $\beta \in \gamma_6$. Similarly, $\alpha \in \gamma_6$. This means that also $\{C_1 C_2 C_1^{-1}, C_1, C_3\}$ is an ordered set.          $\square$

LEMMA 3.2 ([**55**]). *Let*

$$C_1 z = \lambda_1 z \quad (\lambda_1 > 1), \quad C_2(z) = \frac{(\lambda_2 \alpha - \beta)z + (1 - \lambda_2)\alpha\beta}{(\lambda_2 - 1)z + (\alpha - \lambda_2\beta)} \quad (\lambda_2 > 1)$$

$$and \quad C_3 = (C_1 C_2)^{-1}.$$

*Then $\{C_1, C_2, C_3\}$ is a sequential set if and only if*

$$(1) \qquad\qquad 0 < \left( \frac{\sqrt{\lambda_1} + \sqrt{\lambda_2}}{1 + \sqrt{\lambda_1 \lambda_2}} \right)^2 \beta \leqslant \alpha < \beta < \infty.$$

*Here $C_3 \in \mathrm{Aut}_1(\Lambda)$ if and only if*

$$\left( \frac{\sqrt{\lambda_1} + \sqrt{\lambda_2}}{1 + \sqrt{\lambda_1 \lambda_2}} \right)^2 \beta = \alpha.$$

Proof. By the assumptions,

$$C_3^{-1}(z) = \lambda_1 \frac{(\lambda_2\alpha - \beta)z + (1 - \lambda_2)\alpha\beta}{(\lambda_2 - 1)z + (\alpha - \lambda_2\beta)}.$$

The fixed points of $C_3$ are the roots of the equation $C_3^{-1}(x) = x$, which can be rewritten as

(2)    $(\lambda_2 - 1)x^2 - (\lambda_2\beta - \alpha - \lambda_1\beta + \lambda_1\lambda_2\alpha)x + \lambda_1(\lambda_2 - 1)\alpha\beta = 0.$

Therefore, (1) $C_3 \in \mathrm{Aut}_2(H)$ if and only if

$$(\alpha + \lambda_1\lambda_2\alpha - \lambda_1\beta - \lambda_2\beta)^2 - 4\lambda_1\lambda_2(\beta - \alpha)^2 > 0,$$

that is, if

$$\alpha > \Big(\frac{\sqrt{\lambda_1} + \sqrt{\lambda_2}}{1 + \sqrt{\lambda_1\lambda_2}}\Big)^2\beta \quad \text{or} \quad \alpha < \Big(\frac{\sqrt{\lambda_1} - \sqrt{\lambda_2}}{\sqrt{\lambda_1\lambda_2} - 1}\Big)^2\beta;$$

(2) $C_3 \in \mathrm{Aut}_1(H)$ if and only if

$$(\alpha + \lambda_1\lambda_2\alpha - \lambda_1\beta - \lambda_2\beta)^2 - 4\lambda_1\lambda_2(\beta - \alpha)^2 = 0,$$

that is, if

$$\alpha = \Big(\frac{\sqrt{\lambda_1} + \sqrt{\lambda_2}}{1 + \sqrt{\lambda_1\lambda_2}}\Big)^2\beta \quad \text{or} \quad \alpha = \Big(\frac{\sqrt{\lambda_1} - \sqrt{\lambda_2}}{\sqrt{\lambda_1\lambda_2} - 1}\Big)^2\beta.$$

Now let $\{C_1, C_2, C_3\}$ be a sequential set and let $\bar\alpha \leqslant \bar\beta$ be the roots of Eq. (2). Then $0 < \alpha < \beta < \bar\alpha$ (Fig. 1.3.2), whence

$$\frac{\lambda_2\beta - \alpha - \lambda_1\beta + \lambda_1\lambda_2\alpha}{2(\lambda_2 - 1)} = \frac{\bar\alpha + \bar\beta}{2} > \beta.$$

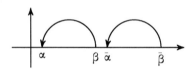

FIGURE 1.3.2

Now, taking into account that $\lambda_1\lambda_2 - 1 > \lambda_1 + \lambda_2 - 2$, we obtain

$$\alpha > \frac{\lambda_1 + \lambda_2 - 2}{\lambda_1\lambda_2 - 1}\beta > \frac{\lambda_1 + \lambda_2 - 2 + 2(1 - \sqrt{\lambda_1\lambda_2})}{\lambda_1\lambda_2 - 1 + 2(1 - \sqrt{\lambda_1\lambda_2})}\beta$$

$$= \frac{\lambda_1 + \lambda_2 - 2\sqrt{\lambda_1\lambda_2}}{\lambda_1\lambda_2 - 2\sqrt{\lambda_1\lambda_2} + 1}\beta = \Big(\frac{\sqrt{\lambda_1} - \sqrt{\lambda_2}}{\sqrt{\lambda_1\lambda_2} - 1}\Big)^2\beta.$$

Therefore,

$$\alpha \geqslant \Big(\frac{\sqrt{\lambda_1} + \sqrt{\lambda_2}}{\sqrt{\lambda_1\lambda_2} + 1}\Big)^2\beta.$$

Now suppose

$$0 < \Big(\frac{\sqrt{\lambda_1} + \sqrt{\lambda_2}}{\sqrt{\lambda_1\lambda_2} + 1}\Big)^2\beta \leqslant \alpha < \beta < \infty$$

and $\bar{\alpha} \leqslant \bar{\beta}$ are the roots of Eq. (2). The inequality $\beta < \bar{\alpha}$ is equivalent to the following pair of relations:

$$(3) \qquad\qquad \bar{\alpha} \cdot \bar{\beta} > \beta^2,$$

$$(4) \qquad (\lambda_2 - 1)\beta^2 - (\lambda_2\beta - \alpha - \lambda_1\beta + \lambda_1\lambda_2\alpha)\beta + \lambda_1(\lambda_2 - 1)\alpha\beta > 0.$$

Equation (3) follows from the obvious inequality

$$\bar{\alpha} \cdot \bar{\beta} = \lambda_1\alpha\beta \geqslant \lambda_1\left(\frac{\sqrt{\lambda_1} + \sqrt{\lambda_2}}{\sqrt{\lambda_1\lambda_2} + 1}\right)^2\beta^2 > \beta^2.$$

Equation (4) is a consequence of the inequality

$$\lambda_2\beta^2 - \beta^2 - \lambda_2\beta^2 + \alpha\beta + \lambda_1\beta^2 - \lambda_1\lambda_2\alpha\beta + \lambda_1\lambda_2\alpha\beta - \lambda_1\alpha\beta$$
$$= (\lambda_1 - 1)(\beta^2 - \alpha\beta) > 0.$$

Hence $\beta < \bar{\alpha}$. The sign of $C_3$ coincides with that of $C_3(0) = C_2^{-1}C_1^{-1}(0) > 0$. Therefore, $\{C_1, C_2, C_3\}$ is an ordered set, which is sequential by Lemma 3.1. $\qquad\square$

LEMMA 3.3. *Let*

$$C_1(z) = \lambda z \quad (\lambda > 1), \quad C_2(z) = \frac{(1 - a\gamma)z + a^2\gamma}{-\gamma z + (1 + a\gamma)} \quad (\gamma > 0)$$
$$and \quad C_3 = (C_1C_2)^{-1}.$$

*Then $\{C_1, C_2, C_3\}$ is a sequential set if and only if $a\gamma \leqslant \frac{\sqrt{\lambda}+1}{\sqrt{\lambda}-1}$. We have $C_3 \in \mathrm{Aut}_1(H)$ if and only if $a\gamma = \frac{\sqrt{\lambda}+1}{\sqrt{\lambda}-1}$.*

Proof. By the assumptions,

$$C_3(z) = \lambda\frac{(1 - a\gamma)z + a^2\gamma}{-\gamma z + (1 + a\gamma)}.$$

The fixed points of $C_3$ are the roots of the equation

$$(5) \qquad\qquad \gamma x^2 - (1 + a\gamma - \lambda + \lambda a\gamma)x + \lambda a^2\gamma = 0.$$

Therefore,

(1) $C_3 \in \mathrm{Aut}_2(H)$ if and only if

$$(1 + a\gamma - \lambda + \lambda a\gamma)^2 > 4\lambda a^2\gamma^2, \quad \text{i.e.,} \quad a\gamma > \frac{\sqrt{\lambda}+1}{\sqrt{\lambda}-1},$$

(2) $C_3 \in \mathrm{Aut}_1(H)$ if and only if

$$(1 + a\gamma - \lambda + \lambda a\gamma)^2 = 4\lambda a^2\gamma^2, \quad \text{i.e.,} \quad a\gamma = \frac{\sqrt{\lambda}+1}{\sqrt{\lambda}-1}.$$

Now suppose that $a\gamma \geqslant \frac{\sqrt{\lambda}+1}{\sqrt{\lambda}-1}$ and let $\bar{\alpha} \leqslant \bar{\beta}$ be the roots of Eq. (5). Then $\bar{\alpha}\bar{\beta} = \lambda a^2 > a^2$ and $\gamma a^2 - (1 + a\gamma - \lambda + \lambda a\gamma)a + \lambda a^2\gamma = a(\lambda - 1) > 0$. Therefore, $0 < a < \bar{\alpha}$ and $\{C_1, C_2, C_3\}$ is an ordered set which is sequential by Lemma 3.1. $\qquad\square$

LEMMA 3.4 ([**52**]). *Let*

$$A(z) = \frac{(\lambda_A\alpha_A - \beta_A)z + (1-\lambda_A)\alpha_A\beta_A}{(\lambda_A - 1)z + (\alpha_A - \lambda_A\beta_A)},$$

$$B(z) = \frac{(\lambda_B\alpha_B - \beta_B)z + (1-\lambda_B)\alpha_B\beta_B}{(\lambda_B - 1)z + (\alpha_B - \lambda_B\beta_B)}$$

*and*

$$C^{-1} = [A, B](z) = \lambda z \quad (\lambda_A, \lambda_B, \lambda > 1).$$

*Then* $\{A, B, C\}$ *is a sequential set of type* $(1,1,0)$ *if and only if*

(6) $$-\infty < \alpha_A < \beta_B < \beta_A < \alpha_B < 0,$$

(7) $$\frac{\alpha_A}{\beta_A} < \sqrt{\lambda}, \quad \frac{\beta_B}{\alpha_B} < \sqrt{\lambda},$$

(8) $$\lambda_A = \frac{\alpha_A\sqrt{\lambda} - \beta_A}{\beta_A\sqrt{\lambda} - \alpha_A}, \quad \lambda_B = \frac{\beta_B\sqrt{\lambda} - \alpha_B}{\alpha_B\sqrt{\lambda} - \beta_B},$$

(9) $$\alpha_B\beta_B\lambda - [(\alpha_A + \beta_A)(\alpha_B + \beta_B) - \alpha_A\beta_A - \alpha_B\beta_B]\sqrt{\lambda} + \alpha_A\beta_A = 0,$$

*and in this case*

$$A(z) = \frac{(\alpha_A + \beta_A)\sqrt{\lambda}z - \alpha_A\beta_A(\sqrt{\lambda} + 1)}{(\sqrt{\lambda} + 1)z - (\alpha_A + \beta_A)},$$

$$B(z) = \frac{(\alpha_B + \beta_B)z - \alpha_B\beta_B(\sqrt{\lambda} + 1)}{(\sqrt{\lambda} + 1)z - (\alpha_B + \beta_B)\sqrt{\lambda}}.$$

Proof. Suppose $\{A, B, C\}$ is a sequential set of type $(1,1,0)$. Then also the set $\{A, BA^{-1}B^{-1}, C\}$ is sequential, whence $-\infty < \alpha_A < \beta_B < \alpha_B < 0$ (see Fig. 1.3.3).

Consider the transformation $\tilde{A} = ABAB^{-1}A^{-1} = AC$. We have

$$\frac{(\lambda_A\alpha_A - \beta_A)(1/\lambda)z + (1-\lambda_A)\alpha_A\beta_A}{(\lambda_A - 1)(1/\lambda)z + (\alpha_A - \lambda_A\beta_A)} = \tilde{A}(z) = \frac{(\lambda_A\tilde{\alpha} - \tilde{\beta})z + (1-\lambda_A)\tilde{\alpha}\tilde{\beta}}{(\lambda_A - 1)z + (\tilde{\alpha} - \lambda_A\tilde{\beta})}$$

(where $\tilde{\alpha} < \tilde{\beta}$ are the fixed points of $\tilde{A}$). Hence

$$\frac{(1-\lambda_A)\alpha_A\beta_A}{(1/\lambda)(\lambda_A - 1)} = \frac{(1-\lambda_A)\tilde{\alpha}\tilde{\beta}}{(\lambda_A - 1)}$$

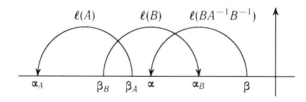

FIGURE 1.3.3

and

$$\frac{(\alpha_A - \beta_A)^2}{(\tilde{\alpha} - \tilde{\beta})^2} = \frac{1}{\lambda}.$$

The first equation implies that $\alpha_A \beta_A = \frac{1}{\lambda} \tilde{\alpha} \tilde{\beta}$ and, therefore,

$$(\tilde{\alpha} + \tilde{\beta})^2 = (\tilde{\alpha} - \tilde{\beta})^2 + 4\tilde{\alpha}\tilde{\beta} = \lambda(\alpha_A + \beta_A)^2.$$

Hence it follows from the equation

$$\tilde{\alpha} + \tilde{\beta} = \frac{(\lambda_A \tilde{\alpha} - \tilde{\beta}) - (\tilde{\alpha} - \lambda_A \tilde{\beta})}{\lambda_A - 1} = \frac{(1/\lambda)(\lambda_A \alpha_A - \beta_A) - (\alpha_A - \lambda_A \beta_A)}{(1/\lambda)(\lambda_A - 1)}$$

that

$$\frac{(\lambda_A \alpha_A - \beta_A) - \lambda(\alpha_A - \lambda_A \beta_A)}{(\lambda_A - 1)} = \sqrt{\lambda}(\alpha_A + \beta_A).$$

Since $\sqrt{\lambda} \neq 1$, we conclude that

$$\sqrt{\lambda} = \frac{\beta_A - \alpha_A \lambda_A}{\alpha_A - \beta_A \lambda_A} > \frac{-\alpha_A \lambda_A}{-\beta_A \lambda_A} = \frac{\alpha_A}{\beta_A},$$

i.e.,

$$\lambda_A = \frac{\alpha_A \sqrt{\lambda} - \beta_A}{\beta_A \sqrt{\lambda} - \alpha_A}.$$

Similarly,

$$\lambda_B = \frac{\beta_B \sqrt{\lambda} - \alpha_B}{\alpha_B \sqrt{\lambda} - \beta_B}$$

and

$$\frac{\beta_B}{\alpha_B} < \sqrt{\lambda}.$$

In the same vein we obtain

$$A(z) = \frac{(\alpha_A + \beta_A)\sqrt{\lambda}z - \alpha_A \beta_A(\sqrt{\lambda} + 1)}{(\sqrt{\lambda} + 1)z - (\alpha_A + \beta_A)},$$

$$B(z) = \frac{(\alpha_B + \beta_B)z - \alpha_B \beta_B(\sqrt{\lambda} + 1)}{(\sqrt{\lambda} + 1)z - (\alpha_B + \beta_B)\sqrt{\lambda}}.$$

The relation $[A, B](z) = \lambda z$ implies Eq. (9).

Now suppose Eqs. (6), (9) are satisfied. Set

$$\tilde{A}(z) = \frac{(\alpha_A + \beta_A)\sqrt{\lambda}z - \alpha_A \beta_A(\sqrt{\lambda} + 1)}{(\sqrt{\lambda} + 1)z - (\alpha_A + \beta_A)},$$

$$\tilde{B}(z) = \frac{(\alpha_B + \beta_B)z - \alpha_B \beta_B(\sqrt{\lambda} + 1)}{(\sqrt{\lambda} + 1)z - (\alpha_B + \beta_B)\sqrt{\lambda}}.$$

Then $\tilde{A}(\alpha_A) = \alpha_A$, $\tilde{A}(\beta_A) = \beta_A$, $\tilde{B}(\alpha_B) = \alpha_B$ and $\tilde{B}(\beta_B) = \beta_B$. By Eq. (9), we have $[\tilde{A}, \tilde{B}] = C^{-1}$. Therefore, $\{\tilde{A}, \tilde{B}\tilde{A}^{-1}\tilde{B}^{-1}, C\}$ is an ordered set which is sequential by Lemma 3.1. Hence $\{\tilde{A}, \tilde{B}, C\}$ is a sequential set of type $(1, 1, 0)$. As has already been proved,

$$\tilde{A}z = \frac{(\tilde{\lambda}_A \alpha_A - \beta_A)z + (1 - \tilde{\lambda}_A)\alpha_A\beta_A}{(\tilde{\lambda}_A - 1)z + (\alpha_A - \lambda_A\beta_A)},$$

where

$$\tilde{\lambda}_A = \frac{\alpha_A\sqrt{\lambda} - \beta_A}{\beta_A\sqrt{\lambda} - \alpha_A} = \lambda_A.$$

Therefore, $\tilde{A} = A$. Similarly, $\tilde{B} = B$.    □

## 4. Frike–Klein–Teichmüller type spaces

**1.** Let $\gamma_{g,n}$ be the group generated by a set of elements

$$v = v_{g,n} = \{a_i, b_i \ (i = 1, \ldots, g), c_i \ (i = g+1, \ldots, n)\}$$

subject to the relation

$$\prod_{i=1}^{g}[a_i, b_i] \prod_{i=g+1}^{n} c_i = 1.$$

Put $c_i = [a_i, b_i] \ (i = 1, \ldots, g)$.

Let $n = g + k + m$ and suppose $6g + 3k + 2m > 6$. Denote by $\tilde{T}_{g,k,m}$ the set of monomorphisms $\psi \colon v \to \mathrm{Aut}(\Lambda)$ such that $\psi(v)$ is a sequential set of type $(g, k, m)$. For $d \in \gamma_{g,n}$ and $D \in \mathrm{Aut}(\Lambda)$ let us set

$$T_{g,k,m}(d, D) = \{\psi \in \tilde{T}_{g,k,m} \mid \psi(d) = D\}.$$

Suppose $k + m > 1$. Consider the embedding

$$\phi \colon \gamma_{g,n-1} \to \gamma_{g,n},$$

uniquely determined by the requirements

$$\phi(a_i) = a_i, \ \phi(b_i) = b_i \qquad\qquad (i = 1, \ldots, g),$$
$$\phi(c_{g+1}) = c_{n-1}c_n, \ \phi(c_i) = c_n^{-1}c_{n-1}^{-1}c_{i-1}c_{n-1}c_n \quad (i = g+2, \ldots, n-1).$$

Consider also the embedding $\tilde{\phi} \colon \gamma_{0,3} \to \gamma_{g,n}$ determined by the conditions

$$\tilde{\phi}(c_1) = c_n^{-1}c_{n-1}^{-1}, \quad \tilde{\phi}(c_2) = c_{n-1}, \quad \tilde{\phi}(c_3) = c_n.$$

A homomorphism $\psi \in \tilde{T}_{g,k,m}$ generates the homomorphisms

$$\Phi(\psi) = \psi\phi, \quad \tilde{\Phi}(\psi) = \psi\tilde{\phi}.$$

For $\psi' \in \Phi(\tilde{T}_{g,k,m})$ denote by $\Phi_{\psi'}$ the restriction of $\tilde{\Phi}$ to $\Phi^{-1}(\psi')$. We set

$$(k', m') = (k+1, m-2), \quad (\tilde{k}, \tilde{m}) = (1, 2) \quad \text{for } m \leqslant 2,$$
$$(k', m') = (k, 0), \qquad\qquad (\tilde{k}, \tilde{m}) = (2, 1) \quad \text{for } m = 1,$$
$$(k', m') = (k-1, 0), \qquad (\tilde{k}, \tilde{m}) = (3, 0) \quad \text{for } m = 0.$$

LEMMA 4.1. $\Phi(\tilde{T}_{g,k,m}) = \tilde{T}_{g,k',m'}$. For any $\psi' \in \tilde{T}_{g,k',m'}$ we have

$$\Phi_{\psi'}(\Phi^{-1}(\psi')) = \tilde{T}_{0,\tilde{k}',\tilde{m}'}(c_1, \psi'(c_{g+1}^{-1})),$$

and the mapping $\Phi_{\psi'}$ is one-to-one.

Proof. Let $\psi \in \tilde{T}_{g,k,m}$ and $\Gamma = \psi(\gamma_{g,n})$. By Theorem 1.1, $\Gamma$ is a Fuchsian group and $P = H/\Gamma$ is a Riemann surface of genus $g$ with $k$ holes and $m$ punctures. According to Lemma 2.1, the set $\psi(v_{g,n})$ generates the standard system of generators

$$\tilde{v} = \{\tilde{a}_i, \tilde{b}_i \ (i = 1, \ldots, g), \tilde{c}_i \ (i = g+1, \ldots, n)\}$$

of $\pi_1(P)$. This implies (see Fig. 1.2.2) that the set

$$\{d_1, \ldots, d_r\} = \{\tilde{a}_1, \tilde{b}_1\tilde{a}_1^{-1}\tilde{b}_1^{-1}, \ldots, \tilde{a}_g, \tilde{b}_g\tilde{a}_g^{-1}\tilde{b}_g^{-1}, \tilde{c}_1, \ldots, \tilde{c}_n\}$$

is such that for all $1 < i < g + n$ the elements

$$d_1 \cdots d_{i-1}, d_i, d_{i+1} \cdots d_r$$

admit a representation by simple contours without intersections. In this case the set

$$\{d_1, \ldots, d_{2g}, d_{r-1}d_r, d_r^{-1}d_{r-1}^{-1}d_{2g+1}d_{r-1}d_r, \ldots, d_r^{-1}d_{r-1}^{-1}d_{r-2}d_{r-1}d_r\}$$

also possesses the same property. By Lemma 2.4, this means that $\psi\phi$ is a sequential set. Hence

$$\Phi(\tilde{T}_{g,k,m}) \subset \tilde{T}_{g,k',m'}.$$

The values $(k', m')$ can easily be found from the geometry of the set $\tilde{v}$ (Fig. 1.2.2).

The fact that

$$\Phi_{\psi'}(\Phi^{-1}(\tilde{T}_{g,k',m'})) \subset \tilde{T}_{0,\tilde{k}',\tilde{m}'}(c_1, \psi'(c_{g+1}^{-1}))$$

is proved similarly.

Now we have to prove that

$$\Phi(\tilde{T}_{g,k,m}) \supset \tilde{T}_{g,k',m'}$$

and to construct, for each $\psi' \in \tilde{T}_{g,k',m'}$, a mapping

$$T_{0,\tilde{k},\tilde{m}}(c_1^{-1}, \psi'(c_{g+1})) \to \Phi^{-1}(\psi'),$$

inverse to $\Phi_{\psi'}$. Let

$$\tilde{\psi} \in T_{0,\tilde{k},\tilde{m}}(c_1^{-1}, \psi'(c_{g+1})).$$

Consider the set of shifts $\psi(v_{g,n})$ defined by the conditions

$$\psi(a_i) = \psi'(a_i), \quad \psi(b_i) = \psi'(b_i) \qquad (i = 1, \ldots, g),$$

$$\psi(c_i) = \psi'(c_{n-1} c_n c_{i-1} c_n^{-1} c_{n-1}^{-1}) \qquad (i = g+2, \ldots, n-2),$$

$$\psi(c_{n-1}) = \tilde{\psi}(c_2), \quad \psi(c_n) = \tilde{\psi}(c_3).$$

The definitions imply that this is a sequential set generating $\psi \in \tilde{T}_{g,k,m}$ and $\Phi(\psi) = \psi'$. Hence $\Phi(\tilde{T}_{g,k,m}) \supset \tilde{T}_{g,k',m'}$. It is easy to see that the correspondence $\tilde{\psi} \mapsto \psi$ is a mapping which is inverse to $\Phi_{\psi'}$.                                    □

**2.**     Suppose $g > 0$. Consider the embedding $\phi^* : \gamma_{g-1,n} \to \gamma_{g,n}$ defined by the conditions $\phi^*(a_i) = a_i$, $\phi^*(b_i) = b_i$ $(i = 1, \ldots, g-1)$, $\phi^*(c_i) = c_i$ $(i = g, \ldots, n)$.

Consider also the embedding $\tilde{\phi}^* : \gamma_{1,1} \to \gamma_{g,n}$ determined by the conditions

$$\tilde{\phi}^*(a_1) = a_g, \quad \tilde{\phi}^*(b_1) = b_g, \quad \tilde{\phi}^*(c_1) = c_g.$$

A homomorphism $\psi \in \tilde{T}_{g,k,m}$ induces the homomorphisms

$$\Omega(\psi) = \psi\phi^*, \quad \tilde{\Omega}(\psi) = \psi\tilde{\phi}^*.$$

For $\psi' \in \Omega(\tilde{T}_{g,k,m})$, denote by $\Omega_{\psi'}$ the restriction of $\tilde{\Omega}$ to $\Omega^{-1}(\psi')$.

LEMMA 4.2. $\Omega(\tilde{T}_{g,k,m}) = \tilde{T}_{g-1,k+1,m}$. For all $\psi' \in \tilde{T}_{g-1,k+1,m}$ we have

$$\Omega_{\psi'}(\Omega^{-1}(\psi')) = \tilde{T}_{1,1,0}(c_1, \psi'(c_g)),$$

and $\Omega_{\psi'}$ is a one-to-one correspondence.

Proof. The inclusion $\Omega(\tilde{T}_{g,k,m}) \subset \tilde{T}_{g-1,k+1,m}$ follows immediately from the definitions. Now we have to prove the opposite inclusion

$$\Omega(\tilde{T}_{g,k,m}) \supset \tilde{T}_{g-1,k+1,m}$$

and construct, for each $\psi' \in \tilde{T}_{g-1,k+1,m}$, a mapping

$$\tilde{T}_{1,1,0}(c_1, \psi'(c_g)) \to \Omega^{-1}(\psi'),$$

which is inverse to $\Omega_{\psi'}$. Take

$$\tilde{\psi} \in \tilde{T}_{1,1,0}(c_1, \psi'(c_g)).$$

Take the set of shifts $\psi(g, n)$ given by the conditions $\psi(a_i) = \psi'(a_i)$, $\psi(b_i) = \psi'(b_i)$ $(i = 1, \ldots, g-1)$, $\psi(a_g) = \tilde{\psi}(a_1)$, $\psi(b_g) = \tilde{\psi}(b_1)$, $\psi(c_i) = \psi'(c_i)$ $(i = g+1, \ldots, n)$. It follows from the definitions that this is a sequential set generating $\psi \in \tilde{T}_{g,k,m}$ and $\Omega(\psi) = \psi'$. Hence

$$\Omega(\tilde{T}_{g,k,m}) = \tilde{T}_{g-1,k+1,m}.$$

It is easy to see that the correspondence $\tilde{\psi} \mapsto \psi$ generates a mapping which is inverse to $\Omega_{\psi'}$.                                    □

**3.**   Now let us parameterize the space $\tilde{T}_{g,k,m}$ following [**55**].

Take $\psi \in \tilde{T}_{g,k,m}$ and set $n = g + k + m$.

We associate to the monomorphism $\psi$ a set of numbers which is for $m = 0$

$$\omega(\psi) = (\lambda_1, \ldots, \lambda_{n-1}, \xi_1, \eta_1, \varepsilon_1, \ldots, \xi_g, \eta_g, \varepsilon_g, \alpha_1, \beta_1, \ldots, \alpha_{g+k-1}, \beta_{g+k-1}),$$

for $m = 1$

$$\omega(\psi) = (\lambda_1, \ldots, \lambda_{n-1}, \xi_1, \eta_1, \varepsilon_1, \ldots, \xi_g, \eta_g, \varepsilon_g, \alpha_1, \beta_1, \ldots,$$
$$\alpha_{g+k-1}, \beta_{g+k-1}, \alpha_{g+k}),$$

and for $m > 1$

$$\omega(\psi) = (\lambda_1, \ldots, \lambda_{n-1}, \xi_1, \eta_1, \varepsilon_1, \ldots, \xi_g, \eta_g, \varepsilon_g, \alpha_1, \beta_1, \ldots,$$
$$\alpha_{g+k}, \beta_{g+k}, a_{g+k+1}, \ldots, a_{n-2}).$$

These numbers are defined by the following conditions:

$$\psi(c_i)(z) = \frac{(\lambda_i \alpha_i - \beta_i)z + (1 - \lambda_i)\alpha_i \beta_i}{(\lambda_i - 1)z + (\alpha_i - \lambda_i \beta_i)}, \quad \lambda_i > 1 \quad \text{for } i \leqslant g + k,$$

$$\psi(c_i)(z) = \frac{(1 - a_i \lambda_i)z + a_i^2 \lambda_i}{-\lambda_i z + (1 + a_i \lambda_i)}, \quad \lambda_i > 0 \quad \text{for } i > g + k,$$

$$\psi(a_i)(z) = \frac{(\lambda_i^a \delta_i - \eta_i)z - (1 - \lambda_i^a)\delta_i \eta_i}{(\lambda_i^a - 1)z + (\delta_i - \lambda_i^a \eta_i)},$$

$$\psi(b_i)(z) = \frac{(\lambda_i^b \varepsilon_i - \xi_i)z - (1 - \lambda_i^b)\varepsilon_i \xi_i}{(\lambda_i^b - 1)z + (\varepsilon_i - \lambda_i^b \xi_i)}.$$

In this way we have associated to the monomorphism $\psi$ a set $\omega(\psi)$ consisting of $6g + 3k + 2m - 3$ numbers.

LEMMA 4.3. *Let* $\psi_1, \psi_2 \in \tilde{T}_{g,k,m}$ *and suppose* $\omega(\psi_1) = \omega(\psi_2)$. *Then* $\psi_1 = \psi_2$.

Proof. By the definitions, the set $\omega(\psi)$ uniquely determines $\psi(c_i)$. Equations (8) and (9) allow one to express $\psi(a_i), \psi(b_i)$ uniquely in terms of $\psi(c_i)$ and $\omega(\psi)$. $\qquad \square$

Hence $\omega(\psi)$ is a system of global coordinates on $\tilde{T}_{g,k,m}$.

The group $\mathrm{Aut}(\Lambda)$ acts on $\tilde{T}_{g,k,m}$ by $\psi \mapsto h^*\psi$, where $\psi \in \tilde{T}_{g,k,m}$, $h \in \mathrm{Aut}(\Lambda)$, $h^*\psi(d) = h\psi(d)h^{-1}$. Set

$$T_{g,k,m} = \tilde{T}_{g,k,m} / \mathrm{Aut}(\Lambda).$$

If $\omega(\psi) = (\lambda_1, \ldots, \lambda_{n-1}, \xi_1, \eta_1, \ldots, a_{n-2})$, then $\omega(h^*\psi) = (\lambda_1, \ldots, \lambda_{n-1}, h\xi_1, h\eta_1, \ldots, ha_{n-2})$.

This property allows us to make the identifications

$$T_{0,0,m} = \{\psi \in \tilde{T}_{0,0,m} \mid \psi(c_1 c_2)(0) = 0, \ \psi(c_1 c_2)(\infty) = \infty, \ a_1 = -1\},$$

$$T_{0,1,m} = \{\psi \in \tilde{T}_{0,1,m} \mid \alpha_1 = \infty, \ \beta_1 = 0, \ a_2 = 1\},$$

$$T_{1,1,m} = \{\psi \in \tilde{T}_{1,1,m} \mid \alpha_1 = \infty, \ \beta_1 = 0, \ \xi_1 = 1\}$$

and in the other cases,

$$T_{g,k,m} = \{\psi \in \tilde{T}_{g,k,m} \mid \alpha_1 = \infty, \ \beta_1 = 0, \ \alpha_2 = 1\}.$$

Now let us describe the mapping
$$\Psi \colon T_{g,k,m} \to \mathbb{R}^{6g+3k+2m-6}.$$
For $g = k = 0$ we set
$$\Psi(\psi) = \{\lambda_1, \lambda_3, \lambda_4, \ldots, \lambda_{m-1}, a_3, \ldots, a_{m-2}\} \subset \mathbb{R}^{2m-6}.$$
For $g = 0, k = 1$ we set
$$\Psi(\psi) = \{\lambda_1, \lambda_2, \ldots, \lambda_m, a_3, \ldots, a_{m-1}\} \subset \mathbb{R}^{2m-3}.$$
For $g = k = 1$ we set
$$\Psi(\psi) = \{\lambda_1, \ldots, \lambda_{m+1}, \eta_1, \varepsilon_1, \alpha_2, \beta_2, a_3, a_4, \ldots, a_m\} \subset \mathbb{R}^{2m+3}.$$
For $g > 1$, $k > 1$, $m = 0$ we set
$$\Psi(\psi) = \{\lambda_1, \ldots, \lambda_{g+k-1}, \xi_1, \eta_1, \varepsilon_1, \ldots, \xi_g, \eta_g, \varepsilon_g,$$
$$\beta_2, \alpha_3, \beta_3, \ldots, \alpha_{g+k-1}, \beta_{g+k-1}\} \subset \mathbb{R}^{6g+3k-6}.$$
For $g > 1$, $k > 1$, $m = 1$ we set
$$\Psi(\psi) = \{\lambda_1, \ldots, \lambda_{g+k}, \xi_1, \eta_1, \varepsilon_1, \ldots, \xi_g, \eta_g, \varepsilon_g, \beta_2,$$
$$\alpha_3, \beta_3, \ldots, \alpha_{g+k-1}, \beta_{g+k-1}, \alpha_{g+k}\} \subset \mathbb{R}^{6g+3k-4}.$$
In the other cases we set
$$\Psi(\psi) = \{\lambda_1, \ldots, \lambda_{g+k+m-1}, \xi_1, \eta_1, \varepsilon_1, \ldots, \xi_g, \eta_g, \varepsilon_g, \beta_2,$$
$$\alpha_3, \beta_3, \ldots, \alpha_{g+k}, \beta_{g+k}, a_{g+k+1}, \ldots, a_{g+k+m-2}\} \subset \mathbb{R}^{6g+3k+2m-6}.$$

Denote the image of $\Psi$ by
$$\hat{T}_{g,k,m} = \Psi(T_{g,k,m}).$$

THEOREM 4.1 ([**55**]). *The mapping $\Psi \colon T_{g,k,m} \to \hat{T}_{g,k,m}$ is a diffeomorphism and $\hat{T}_{g,k,m}$ is an open domain in $\mathbb{R}^{6g+3k+2m-6}$, which is homeomorphic to the space $\mathbb{R}^{6g+3k+2m-6}$ itself.*

Proof. We proceed by induction on $2g + k + m$. For $T_{1,1,0}, T_{0,3,0}, T_{0,2,1}, T_{0,1,2}$ the theorem follows from Lemmas 3.2–3.4. Now let us prove the statement for $T_{g,k,m}$ under the assumption that it is true for all $T_{g',k',m'}$ where $2g' + k' + m' < 2g + k + m$. Suppose $g > 0$. Then, by Lemma 4.2, $\tilde{T}_{g,k,m} = \{\psi\}$ can be represented as a fiber bundle over $\tilde{T}_{g-1,k+1,m}$ whose fiber is homeomorphic to $\tilde{T}_{1,1,0}(c_1^{-1}, \psi(c_g))$. Thus, $T_{g,k,m}$ is represented as a fiber bundle over $T_{g-1,k+1,m}$ with the fiber homeomorphic to
$$\{\psi \in \tilde{T}_{1,1,0} \mid \psi(c_1)(z) = \lambda_g z\}.$$
By Lemma 3.4, the latter set is homeomorphic to $\mathbb{R}^3$. Now the induction hypothesis implies that
$$T_{g,k,m} \cong \mathbb{R}^{6g+3k+2m-6}.$$

Suppose $g = 0$. Then, by Lemma 4.1, $\tilde{T}_{0,k,m} = \{\psi\}$ can be represented as a fiber bundle over $\tilde{T}_{0,k',m'}$ with the fiber homeomorphic to

$$\tilde{T}_{0,\tilde{k},\tilde{m}}(c_1^{-1}, \psi(c_1)).$$

This means that $T_{0,k,m}$ is represented as a fiber bundle over $T_{0,k',m'}$ with the fiber homeomorphic to either $\tilde{T}_{0,3,0}(c_1^{-1}, \psi(c_1))$, or $\tilde{T}_{0,2,1}(c_1^{-1}, \psi(c_1))$, or $\tilde{T}_{0,1,2}(c_1^{-1}, \psi(c_1))$. By Lemmas 3.2 and 3.3, the latter sets are homeomorphic, respectively, to $\mathbb{R}^3, \mathbb{R}^2$, and $\mathbb{R}^1$. It now follows, by the induction hypothesis, that

$$T_{0,k,m} \cong \mathbb{R}^{3k+2m-6}. \qquad \square$$

**Remark.** The defining inequalities for $\hat{T}_{g,k,m}$ in $\mathbb{R}^{6g+3k+2m-6}$ can be made linear or quadratic if we use the variables $\mu_i = \sqrt{\lambda_i}$ instead of $\lambda_i$ .

## 5. Moduli of Riemann surfaces

**1.** Take $\psi \in \tilde{T} = \tilde{T}_{g,k,m}$. By definition, $\psi$ is a mapping $\psi \colon v \to \mathrm{Aut}(\Lambda)$, where $v = v_{g,g+k+m}$ possesses some special properties. Set

$$\widetilde{\mathrm{Mod}}^{\psi} = \widetilde{\mathrm{Mod}}^{\psi}_{g,k,m} = \{\alpha \in \mathrm{Aut}(\gamma) \mid \psi\alpha \in \tilde{T}\},$$

where $\gamma$ is the group generated by $v$.

To the mapping $\psi \in \tilde{T}$, a Fuchsian group $\Gamma^{\psi} = \psi(\gamma)$ and a Riemann surface $P^{\psi} = \Lambda/\Gamma^{\psi}$ are associated. Consider the group $G_p$ of autohomeomorphisms of $P^{\psi}$ preserving a point $p \in P^{\psi}$, the orientation of $P^{\psi}$ and taking holes to holes and punctures to punctures. Consider the subgroup $G'_p \subset G_p$ consisting of homeomorphisms that may be obtained from the identity homeomorphism by a homotopy preserving the point $p$. We set $\widetilde{\mathrm{Mod}}^{\psi}_p = G_p/G'_p$.

LEMMA 5.1. *The groups* $\widetilde{\mathrm{Mod}}^{\psi}$ *and* $\widetilde{\mathrm{Mod}}^{\psi}_p$ *are isomorphic.*

Proof. Consider a uniformization $\Phi \colon \Lambda \to P^{\psi}$. Let $q \in \Phi^{-1}(p)$. Associate to an element $\gamma \in \Gamma^{\psi}$ the element $\Phi(\ell_\gamma) \in \pi_1(P^{\psi}, p)$, where $\ell_\gamma \subset \Lambda$ is a segment connecting $q$ and $\gamma(q)$. The correspondence $\gamma \mapsto \Phi(\ell_\gamma)$ induces an isomorphism $\Phi_q \colon \Gamma^{\psi} \to \pi_1(P^{\psi}, p)$. Now take $\alpha \in \widetilde{\mathrm{Mod}}^{\psi}$. By Lemma 2.1, a standard system of generators of the group $\pi_1(P^{\psi}, p)$ is associated to the sequential set $\psi(v)$. Let us represent these generators by contours that do not intersect each other out of $p$. The preimages of these contours cut from $\Lambda$ a fundamental domain having a vertex at $q$. The sequential set $\psi\alpha(v)$ leads to a similar domain $M^{\alpha}$. Hence there is a homeomorphism $A \colon \Lambda \to \Lambda$ such that $A(M) = M^{\alpha}$ and $A\Gamma^{\psi}A^{-1} = \Gamma^{\psi}$. It induces an autohomeomorphism $\phi_\alpha \in \widetilde{\mathrm{Mod}}^{\psi}_p$ on $P^{\psi}$.

The correspondence $\alpha \mapsto \phi_\alpha$ induces a homomorphism $\phi_q \colon \widetilde{\mathrm{Mod}}^{\psi} \to \widetilde{\mathrm{Mod}}^{\psi}_p$, which is obviously a monomorphism. The fact that it is an epimorphism follows from Theorem 2.1. $\qquad \square$

LEMMA 5.2. *Let* $\psi, \psi' \in \tilde{T}$. *Then* $\widetilde{\mathrm{Mod}}^{\psi} = \widetilde{\mathrm{Mod}}^{\psi'}$.

Proof. Let $\phi_q \colon \widetilde{\mathrm{Mod}}^{\psi} \to \widetilde{\mathrm{Mod}}_p^{\psi}$ and $\phi_q' \colon \widetilde{\mathrm{Mod}}^{\psi'} \to \widetilde{\mathrm{Mod}}_{p'}^{\psi'}$ be the isomorphisms constructed in the course of the proof of Lemma 5.1. Consider the standard systems of generators $v$ and $v'$ of the groups $\pi_1(P^{\psi}, p)$ and $\pi_1(P^{\psi'}, p')$, respectively. The homeomorphism $\beta \colon P \to P'$ taking $v$ to $v'$ induces an isomorphism $\tilde{\beta} \colon \widetilde{\mathrm{Mod}}_p^{\psi} \to \widetilde{\mathrm{Mod}}_{p'}^{\psi'}$. According to our definitions and Lemma 5.1, the element $\phi = \phi_q^{-1} \tilde{\beta}^{-1} \phi_{q'}'$ generates an isomorphism $\phi \colon \widetilde{\mathrm{Mod}}^{\psi'} \to \widetilde{\mathrm{Mod}}^{\psi}$. $\square$

Lemma 5.2 allows us to write $\widetilde{\mathrm{Mod}}$ instead of $\widetilde{\mathrm{Mod}}^{\psi}$.

Each element $h \in \gamma$ generates the automorphism $d \mapsto hdh^{-1}$ of $\gamma$, which belongs to $\widetilde{\mathrm{Mod}}$. Denote by $\widetilde{\mathrm{IMod}} \subset \widetilde{\mathrm{Mod}}$ the subgroup of all such diffeomorphisms and set $\mathrm{Mod} = \widetilde{\mathrm{Mod}}/\widetilde{\mathrm{IMod}}$.

Similarly, each element $h \in \pi(P^{\psi}, p)$ generates the automorphism $d \mapsto hdh^{-1}$ of $\pi_1(P^{\psi}, p)$, which belongs to $\widetilde{\mathrm{Mod}}_p^{\psi}$. We denote by $\widetilde{\mathrm{IMod}}_p^{\psi} \subset \widetilde{\mathrm{Mod}}_p^{\psi}$ the subgroup of all such diffeomorphisms and set $\mathrm{Mod}_p^{\psi} = \widetilde{\mathrm{Mod}}_p^{\psi}/\widetilde{\mathrm{IMod}}_p^{\psi}$. By moving the point $p$ we obtain natural isomorphisms between the groups $\mathrm{Mod}_p^{\psi}$ corresponding to different $p$'s. Hence, each of these groups is naturally isomorphic to $\mathrm{Mod}_*^{\psi} = \widetilde{\mathrm{Mod}}_*^{\psi}/\widetilde{\mathrm{IMod}}_*^{\psi}$, where $\widetilde{\mathrm{Mod}}_*^{\psi}$ is the group of orientation preserving autohomeomorphisms of $P^{\psi}$ taking holes to holes and punctures to punctures, and $\widetilde{\mathrm{IMod}}_*^{\psi} \subset \widetilde{\mathrm{Mod}}_*^{\psi}$ is the subgroup of homeomorphisms homotopic to the identity. Homeomorphisms $P^{\psi} \to P^{\psi'}$ taking holes to holes and punctures to punctures induce natural automorphisms $\widetilde{\mathrm{Mod}}_*^{\psi} \to \widetilde{\mathrm{Mod}}_*^{\psi'}$. This property allows us to omit the indices $\psi$ in what follows and use the notation $\widetilde{\mathrm{Mod}}_* = \widetilde{\mathrm{Mod}}_*^{\psi}$, $\mathrm{Mod}_* = \mathrm{Mod}_*^{\psi}$. Lemma 5.1 implies

COROLLARY 5.1. *The group* $\mathrm{Mod} = \mathrm{Mod}_{g,k,m}$ *is naturally isomorphic to the group* $\mathrm{Mod}_* = \mathrm{Mod}_*^{g,k,m}$ *of homotopy classes of orientation preserving autohomeomorphisms of the surfaces of type* $(g, k, m)$.

**2.** Now let us describe the moduli spaces (that is, spaces of classes of biholomorphic equivalence) of Riemann surfaces.

THEOREM 5.1. *The group* $\mathrm{Mod}_{g,k,m}$ *naturally acts on* $T_{g,k,m}$ *by diffeomorphisms. This action is discrete. The quotient space* $T_{g,k,m}/\mathrm{Mod}_{g,k,m}$ *is naturally identified with the moduli space* $M_{g,k,m}$ *of Riemann surfaces of type* $(g, k, m)$.

Proof. By Lemma 5.2, the group $\widetilde{\mathrm{Mod}} = \widetilde{\mathrm{Mod}}_{g,k,m}$ acts naturally on the space $\tilde{T} = \tilde{T}_{g,k,m}$. The group $\mathrm{Aut}(\Lambda)$ also acts on the same space. It is easy to see that $\widetilde{\mathrm{Mod}} \cap \mathrm{Aut}(\Lambda) = \widetilde{\mathrm{IMod}}$. Thus, the group Mod acts naturally on $T = T_{g,k,m}$. The coordinates on $T_{g,k,m} = \{\psi\}$ introduced in Section 4 give an explicit analytic description of a sequential set $\psi(v)$ as a subset in $(\mathrm{PSL}(2,\mathbb{R}))^{2g+k+m-1}$. Therefore, the analyticity of the action of Mod on $T$ follows from the analyticity of the group operation in $\mathrm{PSL}(2,\mathbb{R})$.

Consider the mapping $\tilde{\Phi}\colon \tilde{T} \to M = M_{g,k,m}$, where $\tilde{\Phi}(\psi) = P^{\psi}$. It is clear that $\tilde{\Phi}$ takes coinciding values on the orbits of both $\mathrm{Aut}(\Lambda)$ and $\widetilde{\mathrm{Mod}}$. Hence $\tilde{\Phi}$ induces a mapping $\Phi\colon T/\mathrm{Mod} \to M$. By the uniformization theorem, each surface $P \in M$ has a representation of the form $\Lambda/\Gamma$, where $\Gamma$ is a Fuchsian group. Now Theorem 2.1 implies that $\Phi(T) = \tilde{\Phi}(\tilde{T}) = M$.

Now let us prove that $\Phi(x) \neq \Phi(x')$ for $x \neq x'$. Indeed, suppose $\psi \in x$, $\psi' \in x'$ and $\tilde{\Phi}(\psi) = \tilde{\Phi}(\psi')$. This means, by definition, that the Riemann surfaces $P^{\psi}$ and $P^{\psi'}$ are biholomorphically equivalent. Therefore, $\psi$ and $\psi'$ can be chosen so that the uniformizing groups $\Gamma^{\psi}$ and $\Gamma^{\psi'}$ coincide. Hence there is an element $\alpha \in \widetilde{\mathrm{Mod}}^{\psi}$ such that $\psi' = \psi\alpha$. By Lemma 5.2, we have $\alpha \in \widetilde{\mathrm{Mod}}$, whence $x = x'$.

To each element $\psi \in \tilde{T}$ we associate the following set of lengths of geodesics on $P^{\psi}$. The length of the geodesic corresponding to $C \in \Gamma^{\psi}$ is $\ln(1/4)(\lambda + 1/\lambda + \sqrt{\lambda + 1/\lambda - 2})$, where $\lambda$ is the shift parameter of $C$ [**94**]. We conclude that Mod acts discretely, since the set of the lengths of the geodesics is discrete [**94**]. $\qquad\square$

By Theorem 5.1, the analytic structure on $T$ introduced in Section 4 induces an analytic structure on the moduli space $M$. This is exactly the analytic structure we use in our study of the topology of $M$. Theorems 4.1 and 5.1 give

COROLLARY 5.2 ([**28**]). *The moduli space of Riemann surfaces of type* $(g, k, m)$ *is diffeomorphic to* $\mathbb{R}^{6g+3k+2m-6}/\mathrm{Mod}_{g,k,m}$, *where* $\mathrm{Mod}_{g,k,m}$ *is a discrete group.*

## 6. The space of holomorphic morphisms of Riemann surfaces

**1.** A holomorphic morphism of degree $d$ of Riemann surfaces is a triple $(\tilde{P}, f, P)$, where $\tilde{P}$ and $P$ are Riemann surfaces and $f\colon \tilde{P} \to P$ is a holomorphic mapping of degree $d$ such that $f(\tilde{P}) = P$. Two morphisms $(\tilde{P}_1, f_1, P_1)$ and $(\tilde{P}_2, f_2, P_2)$ are considered the same if there are biholomorphic mappings $\tilde{\phi}\colon \tilde{P}_1 \to \tilde{P}_2$ and $\phi\colon P_1 \to P_2$ such that $\phi f_1 = f_2 \tilde{\phi}$.

Two morphisms $(\tilde{P}_1, f_1, P_1)$ and $(\tilde{P}_2, f_2, P_2)$ are said to be *topologically equivalent* if there are homeomorphisms $\tilde{\phi}\colon \tilde{P}_1 \to \tilde{P}_2$ and $\phi\colon P_1 \to P_2$ such that $\phi f_1 = f_2 \tilde{\phi}$. A class of topological equivalence of morphisms is called a *topological type*. Fix a topological type $t$ and consider the set $H^t$ of all morphisms of type $t$.

EXAMPLE. If $P$ is the Riemann sphere, then $(\tilde{P}, f, P)$ is a meromorphic function on $\tilde{P}$. If $\tilde{P}$ also is the Riemann sphere, then $f$ is a rational function. If, moreover, $f^{-1}(\infty) = \infty$, then $f$ is a polynomial. A *critical value* of a polynomial is its value $f(x)$ at a point $x$ where $df(x) = 0$. If all critical values are distinct, then we say that the polynomial is *general*. All general polynomials are topologically equivalent. However, generally speaking, the behavior of the critical values does not determine the topological type of a polynomial uniquely [96]. Other examples of the sets $H^t$ include general Laurent polynomials [3], and spaces of meromorphic functions with fixed orders of poles [59], [63], [75], which arise in topological field theories [19].

The topological type of a morphism $(\tilde{P}, f, P) \in H^t$ uniquely determines the topological type $(g, k, m)$ of the surface $P$ it covers. Besides, $t$ uniquely determines the number $b$ of the critical values

$$B = B(\tilde{P}, f, P) = \{ f(\tilde{p}) \in P \mid \tilde{p} \in \tilde{P}, df(\tilde{p}) = 0 \}.$$

We assume that $b < \infty$. Then the correspondence $(\tilde{P}, f, P) \mapsto P \setminus B$ defines a mapping

$$\Psi \colon H^t \to M_{g,k,m+b}.$$

In Section 5 we have introduced a topology in $M_{g,k,m+b}$. Now let us introduce a topology in $H^t$ by declaring the preimages $\Psi^{-1}(U)$ of open sets $U \subset M_{g,k,m+b}$ to be open sets in $H^t$.

**2.** Now take $\psi \in \tilde{T}_{g,k,m+b}$ and let $\tilde{\gamma}$ be a subgroup of index $d$ of the group $\gamma = \gamma_{g,n}$, where $n = g + k + m + b$. Set

$$\Gamma^\psi = \psi(\gamma), \quad \Gamma^\psi_{\tilde{\gamma}} = \psi(\tilde{\gamma}), \quad P^\psi = \Lambda/\Gamma^\psi \quad \text{and} \quad P^\psi_{\tilde{\gamma}} = \Lambda/\Gamma^\psi_{\tilde{\gamma}}.$$

Then the natural embedding $\Gamma^\psi_{\tilde{\gamma}} \subset \Gamma^\psi$ induces a morphism

$$f^\psi_{\tilde{\gamma}} \colon P^\psi_{\tilde{\gamma}} \to P^\psi$$

of degree $d$. After patching on $P^\psi$ and $P^\psi_{\tilde{\gamma}}$ the punctures generated by the parabolic shifts $\psi(c_{g+k+m+1}), \ldots, \psi(c_n)$, we obtain a morphism

$$(P^\psi_{\tilde{\gamma}}, f^\psi_{\tilde{\gamma}}, P^\psi)^b.$$

LEMMA 6.1. *Each holomorphic morphism has the form*

$$(P^\psi_{\tilde{\gamma}}, f^\psi_{\tilde{\gamma}}, P^\psi)^b,$$

*where $\psi \in \tilde{T}_{g,k,m+b}$ and $\tilde{\gamma} \subset \gamma_{g,n}$.*

Proof. Let $(\tilde{P}, f, P)$ be an arbitrary holomorphic morphism. Set $P^* = P \setminus B(\tilde{P}, f, P) \in M_{g,k,m+b}$ and $\tilde{P}^* = f^{-1}(P^*)$. By Theorem 5.1, there is a morphism $\psi \in \tilde{T}_{g,k,m+b}$ such that $P^\psi = P^*$. Theorem 2.1 implies that $\psi(c_{g+k+m+1}), \ldots, \psi(c_n)$ correspond to punctures in $B = B(\tilde{P}, f, P) \subset P$.

The group $\Gamma^\psi = \psi(\gamma)$ is the monodromy group of the universal covering $\Phi \colon \Lambda \to P^*$. Because of the universality, the covering $\Phi$ can be factored through a covering of $P^*$ and, in particular, there is a universal covering $\tilde{\Phi} \colon \Lambda \to \tilde{P}^*$ such that $\Phi = f\tilde{\Phi}$. Let $\tilde{\Gamma}^\psi \subset \Gamma^\psi$ be the monodromy group of this covering. Set $\tilde{\gamma} = \psi^{-1}(\tilde{\Gamma}^\psi)$. Then

$$(P_{\tilde{\gamma}}^\psi, f_{\tilde{\gamma}}^\psi, P^\psi) = (\tilde{P}^*, f^*, P^*),$$

where $f^* = f|_{\tilde{P}^*}$. Patching up the punctures in $B$ we arrive at the covering

$$(P_{\tilde{\gamma}}^*, f_{\tilde{\gamma}}^\psi, P^\psi)^b = (\tilde{P}, f, P).$$

$\square$

LEMMA 6.2. *Let* $\psi, \psi' \in \tilde{T}_{g,k,m+b}$ *and let* $\tilde{\gamma} \subset \gamma_{g,n}$ *be a subgroup. Then* $(P_{\tilde{\gamma}}^\psi, f_{\tilde{\gamma}}^\psi, P^\psi)^b$ *and* $(P_{\tilde{\gamma}}^{\psi'}, f_{\tilde{\gamma}}^{\psi'}, P^{\psi'})^b$ *have the same topological type. If, in addition, there is* $A \in \mathrm{Aut}(\Lambda)$ *such that* $A\psi(w)A^{-1} = \psi'(w)$ *for all* $w \in v_{g,n}$, *then*

$$(P_{\tilde{\gamma}}^\psi, f_{\tilde{\gamma}}^\psi, P^\psi)^b = (P_{\tilde{\gamma}}^{\psi'}, f_{\tilde{\gamma}}^{\psi'}, P^{\psi'})^b.$$

Proof. Set $v = v_{g,g+k+m+b}$. By Lemma 2.1, the natural coverings $\Phi \colon \Lambda \to P^\psi$, $\Phi' \colon \Lambda \to P^{\psi'}$ take the sets $\psi(v)$ and $\psi'(v)$ to standard systems of generators of the groups $\pi_1(P^\psi, p)$ and $\pi_1(P^{\psi'}, p')$. Therefore, there is a homeomorphism $\phi \colon P^\psi \to P^{\psi'}$ taking the first standard system to the second one. The projections $\Phi_q$ and $\Phi'_q$ lift this homeomorphism to a homeomorphism $A \colon \Lambda \to \Lambda$ such that $A\psi(w)A^{-1} = \psi'(w)$ for any $w \in v$. Then it follows that $A\psi(\tilde{\gamma})A^{-1} = \psi'(\tilde{\gamma})$. This means that $A$ induces a homeomorphism $\tilde{\phi} \colon \tilde{P}^\psi \to \tilde{P}^{\psi'}$ such that $\phi f = f'\tilde{\phi}$. If, in addition, $A \in \mathrm{Aut}(\Lambda)$, then $\phi$ and $\tilde{\phi}$ are holomorphic mappings and, therefore,

$$(P_{\tilde{\gamma}}^\psi, f_{\tilde{\gamma}}^\psi, P^\psi)^b = (P_{\tilde{\gamma}}^{\psi'}, f_{\tilde{\gamma}}^{\psi'}, P^{\psi'})^b.$$

$\square$

In the group $\gamma = \gamma_{g,n}$ generated by the elements

$$v_{g,n} = \{a_i, b_i \ (i = 1, \ldots, g), c_i \ (i = g+1, \ldots, n)\},$$

we choose the subset $c$ of elements of the form $hc_ih^{-1}$, where $i > g+k+m$, $h \in \gamma$.

Set

$$\widetilde{\mathrm{Mod}}_{g,k,m+b}^b = \{\alpha \in \widetilde{\mathrm{Mod}}_{g,k,m+b} \mid \alpha(c) = c\}.$$

LEMMA 6.3. *Two morphisms* $(P_{\tilde{\gamma}}^\psi, f_{\tilde{\gamma}}^\psi, P^\psi)^b$ *and* $(P_{\gamma'}^\psi, f_{\gamma'}^\psi, P^\psi)^b$ *have the same topological type if and only if* $\gamma' = \alpha(\tilde{\gamma})$, *where* $\alpha \in \widetilde{\mathrm{Mod}}_{g,k,m+b}^b$.

Proof. Let $\tilde{\phi} \colon P_{\tilde{\gamma}}^\psi \to P_{\gamma'}^\psi$ and $\phi \colon P^\psi \to P^\psi$ be homeomorphisms establishing topological equivalence between $(P_{\tilde{\gamma}}^\psi, f_{\tilde{\gamma}}^\psi, P^\psi)^b$ and $(P_{\gamma'}^\psi, f_{\gamma'}^\psi, P^{\psi'})^b$. The natural coverings $\Phi \colon \Lambda \to P_{\tilde{\gamma}}^\psi \to P^\psi$ and $\Phi' \colon \Lambda \to P_{\gamma'}^\psi \to P^\psi$ lift them to a homeomorphism $A \colon \Lambda \to \Lambda$ such that $\psi(\gamma) = A\psi(\gamma)A^{-1}$ and $\psi(\gamma') =$

$A\psi(\tilde{\gamma})A^{-1}$. Denote by $A^*\colon \psi(\gamma) \to \psi(\gamma)$ the homeomorphism $A^*(h) = AhA^{-1}$ and set $\alpha = \psi^{-1}A^*\psi\colon \gamma \to \gamma$. Then $\alpha \in \widetilde{\mathrm{Mod}}^b_{g,k,m+b}$ and $\alpha(\tilde{\gamma}) = \gamma'$.

Now let $\gamma' = \alpha(\tilde{\gamma})$, where $\alpha \in \widetilde{\mathrm{Mod}}^b_{g,k,m+b}$. Set $\psi' = \psi\alpha \in \tilde{T}_{g,k,m+b}$. Then

$$(P^{\psi'}_{\tilde{\gamma}}, f^{\psi'}_{\tilde{\gamma}}, P^{\psi'})^b = (P^{\psi}_{\gamma'}, f^{\psi}_{\gamma'}, P^{\psi})^b,$$

and, by Lemma 6.2, $(P^{\psi'}_{\tilde{\gamma}}, f^{\psi'}_{\tilde{\gamma}}, P^{\psi'})^b$ and $(P^{\psi}_{\tilde{\gamma}}, f^{\psi}_{\tilde{\gamma}}, P^{\psi})^b$ have the same topological type. $\qquad\square$

LEMMA 6.4. *Take* $(P^{\psi}_{\tilde{\gamma}}, f^{\psi}_{\tilde{\gamma}}, P^{\psi}) \in H^t$. *Then*

$$H^t = \{(P^{\psi'}_{\tilde{\gamma}}, f^{\psi'}_{\tilde{\gamma}}, P^{\psi'})^b \mid \psi' \in \tilde{T}_{g,k,m+b}\}.$$

Proof. By Lemma 6.2, we have

$$\{(P^{\psi'}_{\tilde{\gamma}}, f^{\psi'}_{\tilde{\gamma}}, P^{\psi'})^b \mid \psi' \in \tilde{T}_{g,k,m+b}\} \subset H^t.$$

Let us prove the inverse inclusion. Take $(\tilde{P}, f, P) \in H^t$. By Lemma 6.1, we have

$$(\tilde{P}, f, P) = (P^{\psi'}_{\gamma'}, f^{\psi'}_{\gamma'}, P^{\psi'})^b,$$

where, by Lemma 6.3, $\gamma' = \alpha(\tilde{\gamma})$ for $\alpha \in \widetilde{\mathrm{Mod}}^b_{g,k,m+b}$. Set $\psi^* = \psi'\alpha$. Then

$$(P^{\psi'}_{\gamma'}, f^{\psi'}_{\gamma'}, P^{\psi'})^b = (P^{\psi^*}_{\tilde{\gamma}}, f^{\psi^*}_{\tilde{\gamma}}, P^{\psi^*})^b$$

and $\psi^* \in \tilde{T}_{g,k,m+b}$. $\qquad\square$

**3.** For a subgroup $\tilde{\gamma} \subset \gamma = \gamma_{g,n}$, we set

$$\widetilde{\mathrm{Mod}}^{\tilde{\gamma}}_{g,k,m,b} = \{\alpha \in \widetilde{\mathrm{Mod}}^b_{g,k,m+b} \mid \alpha\tilde{\gamma} = \tilde{\gamma}\},$$

$$\widetilde{\mathrm{IMod}}^{\tilde{\gamma}}_{g,k,m,b} = \widetilde{\mathrm{Mod}}^{\tilde{\gamma}}_{g,k,m,b} \cap \widetilde{\mathrm{IMod}}_{g,k,m+b}$$

and

$$\mathrm{Mod}^{\tilde{\gamma}}_{g,k,m,b} = \widetilde{\mathrm{Mod}}^{\tilde{\gamma}}_{g,k,m,b} \setminus \widetilde{\mathrm{IMod}}^{\tilde{\gamma}}_{g,k,m,b}.$$

An automorphism $\alpha \in \widetilde{\mathrm{IMod}}^{\tilde{\gamma}}_{g,k,m,b}$ takes an element $\psi \in \tilde{T}_{g,k,m+b}$ to $h^*\psi \in \tilde{T}_{g,k,m+b}$, where $h^*\psi(w) = h\psi(w)h^{-1}$ and $h \in \mathrm{Aut}(\Lambda)$. Hence the group $\mathrm{Mod}^{\tilde{\gamma}}_{g,k,m,b}$ acts on $T_{g,k,m+b}$.

Let $N(g,n,d)$ denote the set of subgroups of index $d$ in $\gamma_{g,n}$. The group $\mathrm{Mod}^b_{g,k,m+b}$ acts on this set. Denote by $n(g,k,m,b,d)$ the number of orbits of this action.

Let $H^{b,d}_{g,k,m}$ denote the space of holomorphic morphisms $(\tilde{P}, f, P)$ such that $P \in H_{g,k,m}$, the degree of $f$ is $d$ and the number of critical values of $f$ is $b$.

THEOREM 6.1. *The space* $H = H^{b,d}_{g,k,m}$ *decomposes into* $n(g,k,m,b,d)$ *connected components. Each of these connected components is a class of topological equivalence of holomorphic morphisms and is homeomorphic to* $T_{g,k,m+b}/\widetilde{\mathrm{Mod}}^{\tilde{\gamma}}_{g,k,m,b}$, *where the group* $\mathrm{Mod}^{\tilde{\gamma}}_{g,k,m,b}$ *acts discretely.*

Proof. According to Lemmas 6.4 and 6.3, each connected component of $H$ has the form

$$H^t = \{(P_{\tilde{\gamma}}^{\psi}, f_{\tilde{\gamma}}^{\psi}, P^{\psi})^b \mid \psi \in \tilde{T}_{g,k,m+b}\}$$

and the number of such connected components is $n(g, k, m, b, d)$. Consider the mapping $\tilde{\Phi} \colon \tilde{T}_{g,k,m+b} \to H^t$, where $\tilde{\Phi}(\psi) = (P_{\tilde{\gamma}}^{\psi}, f_{\tilde{\gamma}}^{\psi}, P^{\psi})^b$. By Lemma 6.2, this mapping induces a mapping $\Phi \colon T_{g,k,m+b} \to H^t$. It is easy to see that a fiber of $\Phi$ is an orbit of $\mathrm{Mod}_{g,k,m,b}^{\tilde{\gamma}}$. The action of $\mathrm{Mod}_{g,k,m,b}^{\tilde{\gamma}}$ is discrete, since the action of $\mathrm{Mod}_{g,k,m,b}$ on $T_{g,k,m+b}$ is discrete (Theorem 5.1). $\qquad\square$

COROLLARY 6.1. *The space of all holomorphic morphisms of a given topological type is connected and homeomorphic to* $\mathbb{R}^s / \mathrm{Mod}$, *where* $\mathrm{Mod}$ *is a discrete group.*

**Remark.** Some special cases of this theorem where known before: (1) for general Laurent polynomials [3]; (2) for general meromorphic functions (together with a description of Mod) [60]; (3) for $g = 0$ [83]; (4) for compact surfaces [84].

## 7. Lifting Fuchsian groups to $\mathrm{SL}(2, \mathbb{R})$

**1.** The solution of the following problem will play an important role in our study of spinor bundles and super-Riemann surfaces. Let

$$\Gamma \subset \mathrm{Aut}_1(\Lambda) \cup \mathrm{Aut}_2(\Lambda) \subset \mathrm{Aut}(\Lambda) = \mathrm{PSL}(2, \mathbb{R})$$

be a finitely generated Fuchsian group. We want to describe all its liftings to $\mathrm{SL}(2, \mathbb{R})$, that is, the subgroups $\Gamma^* \subset \mathrm{SL}(2, \mathbb{R})$ the restrictions to which of the natural projection $J \colon \mathrm{SL}(2, \mathbb{R}) \to \mathrm{PSL}(2, \mathbb{R})$ induce an isomorphism $\Gamma^* \to \Gamma$.

To start, let us set

$$\mathrm{Aut}_i^* = J^{-1}(\mathrm{Aut}_i(\Lambda)) \subset \mathrm{SL}(2, \mathbb{R}) \quad (i = 1, 2)$$

and associate to each matrix $C \in \mathrm{Aut}_i^*$ a number $\sigma(C) \in \{1, -1\}$ uniquely determined by the following condition: the matrix $C$ is adjacent either to the matrix

$$\sigma(C) \begin{pmatrix} \lambda & 0 \\ 0 & \lambda^{-1} \end{pmatrix},$$

where $\lambda > 0$ (if $C \in \mathrm{Aut}_2^*$), or to the matrix

$$\sigma(C) \begin{pmatrix} 1 & 1 \\ 0 & 1 \end{pmatrix}$$

(if $C \in \mathrm{Aut}_1^*$). Thus, the sign of $\sigma(C)$ coincides with that of the trace of $C$.

LEMMA 7.1. *Let* $C_1, C_2, C_3 \in \mathrm{SL}(2, \mathbb{R})$ *be such that* $\{J(C_1), J(C_2), J(C_3)\}$ *is a sequential set of type* $(0, k, m)$. *Then* $C_1 C_2 C_3 = 1$ *if and only if*

$$\sigma(C_1)\sigma(C_2)\sigma(C_3) = -1.$$

Proof. Suppose first that $C_1, C_2 \in \mathrm{Aut}_2^*$. After conjugating, if necessary, the pair $(C_1, C_2)$ by a matrix in $\mathrm{SL}(2, \mathbb{R})$ we may assume that

$$C_1 = \frac{\sigma(C_1)}{\sqrt{\lambda_1}} \begin{pmatrix} \lambda_1 & 0 \\ 0 & 1 \end{pmatrix}$$

and

$$C_2 = \frac{\sigma(C_2)}{\sqrt{\lambda_2}(\alpha - \beta)} \begin{pmatrix} (\lambda_2\alpha - \beta) & (1 - \lambda_2)\alpha\beta \\ (\lambda_2 - 1) & (\alpha - \lambda_2\beta) \end{pmatrix},$$

where $\lambda_1, \lambda_2 > 1, 0 < \alpha < \beta$. Therefore,

$$C_1 C_2 = \frac{\sigma(C_1)\sigma(C_2)}{\sqrt{\lambda_1\lambda_2}(\alpha - \beta)} \begin{pmatrix} \lambda_1(\lambda_2\alpha - \beta) & \lambda_1(1 - \lambda_2)\alpha\beta \\ (\lambda_2 - 1) & (\alpha - \lambda_2\beta) \end{pmatrix}.$$

This matrix has the characteristic equation

$$x^2 - (-\lambda_2\beta + \alpha - \lambda_1\beta + \lambda_1\lambda_2\alpha)x + \lambda_1(\lambda_1 - 1)(\lambda_2 - 1)\alpha\beta = 0$$

whose roots have the same sign coinciding with that of

$$-\lambda_2\beta + \alpha - \lambda_1\beta + \lambda_1\lambda_2\alpha.$$

By Lemma 3.2, we have

$$0 < \left(\frac{\sqrt{\lambda_1} + \sqrt{\lambda_2}}{1 + \sqrt{\lambda_1\lambda_2}}\right)^2 \beta \leqslant \alpha < \beta < \infty.$$

Therefore,

$$\alpha \geqslant \left(\frac{\sqrt{\lambda_1} + \sqrt{\lambda_2}}{\sqrt{\lambda_1\lambda_2} + 1}\right)^2 \beta \geqslant \frac{\lambda_1 + \lambda_2}{\lambda_1\lambda_2 + 1},$$

whence

$$-\lambda_2\beta + \alpha - \lambda_1\beta + \lambda_1\lambda_2\alpha > 0.$$

Thus,

$$\sigma(C_1 C_2) = \sigma(C_1)\sigma(C_2)\,\mathrm{sign}(\alpha - \beta) = -\sigma(C_1)\sigma(C_2),$$

and the condition $C_3^{-1} = C_1 C_2$ is equivalent to

$$\sigma(C_3) = \sigma(C_3^{-1}) = \sigma(C_1 C_2) = -\sigma(C_1)\sigma(C_2).$$

Similarly, if $C_1 \in \mathrm{Aut}_2^*$, $C_2 \in \mathrm{Aut}_1^*$, then after conjugating, if necessary, by an element in $\mathrm{SL}(2, \mathbb{R})$, we may assume that

$$C_1 = \frac{\sigma(C_1)}{\sqrt{\lambda}} \begin{pmatrix} \lambda & 0 \\ 0 & 1 \end{pmatrix}$$

and

$$C_2 = \sigma(C_2) \begin{pmatrix} (1 - \alpha\gamma) & a^2\gamma \\ -\gamma & (1 + a\gamma) \end{pmatrix},$$

where $\lambda > 1, \gamma > 0$. Therefore,

$$C_1 C_2 = \frac{\sigma(C_1)\sigma(C_2)}{\sqrt{\lambda}} \begin{pmatrix} \lambda(1 - a\gamma) & \lambda a^2\gamma \\ -\gamma & (1 + a\gamma) \end{pmatrix}.$$

The characteristic equation of this matrix is

$$x^2 - [\lambda(1 - a\gamma) + (1 + a\gamma)]x + \lambda a^2\gamma^2 = 0.$$

The roots of this equation have the same sign coinciding with that of $\lambda(1 - a\gamma) + (1 + a\gamma)$. By Lemma 3.3,

$$a\gamma \geqslant \frac{\sqrt{\lambda} + 1}{\sqrt{\lambda} - 1} = \frac{(\sqrt{\lambda} + 1)^2}{\lambda - 1},$$

whence

$$\lambda(1 - a\gamma) + (1 + a\gamma) = (1 + \lambda) + a\gamma(1 - \lambda) \leqslant (1 + \lambda) - (\sqrt{\lambda} + 1)^2 < 0.$$

Thus,

$$\sigma(C_1 C_2) = -\sigma(C_1)\sigma(C_2),$$

and the condition $C_3^{-1} = C_1 C_2$ is equivalent to

$$\sigma_1(C_1)\sigma(C_2)\sigma(C_3) = -1.$$

The above argument proves the lemma in the case $k > 0$. If $k = 0$, then, after a conjugation, we may assume that

$$C_1 = \sigma(C_1)\begin{pmatrix} 1 & 1 \\ 0 & 1 \end{pmatrix}, \quad C_2 = \sigma(C_2)\begin{pmatrix} -3 & 4 \\ -4 & 5 \end{pmatrix},$$

whence

$$\sigma(C_1 C_2) = -\sigma(C_1)\sigma(C_2). \qquad \square$$

LEMMA 7.2. *Let* $A, B, C \in SL(2, \mathbb{R})$ *be such that* $\{J(A), J(B), J(C)\}$ *is a sequential set of type* $(1, 1, 0)$. *Then* $[A, B]C = 1$ *if and only if* $\sigma(C) = -1$. *Moreover,* $\sigma(AB) = \sigma(A)\sigma(B)$.

Proof. By definition, $\sigma(A) = \sigma(BA^{-1}B^{-1})$ and $\{J(A), J(BA^{-1}B^{-1}), J(C)\}$ is a sequential set of type $(0, 3, 0)$. Hence the first assertion of the lemma follows from Lemma 7.1. Let us prove that $\sigma(AB) = \sigma(A)\sigma(B)$. After conjugating by an element in $SL(2, \mathbb{R})$ we may assume that

$$A = \frac{\sigma(A)}{\sqrt{\lambda_A}}\begin{pmatrix} \lambda & 0 \\ 0 & 1 \end{pmatrix}$$

and

$$B = \frac{\sigma(B)}{\sqrt{\lambda_2}(\alpha - \beta)}\begin{pmatrix} (\lambda_2\alpha - \beta) & (1 - \lambda_2)\alpha\beta \\ (\lambda_2 - 1) & (\alpha - \lambda_2\beta) \end{pmatrix},$$

where $\lambda_1, \lambda_2 > 1$, $\beta < 0 < \alpha$ (see Fig. 1.7.1).

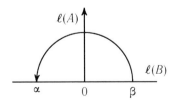

FIGURE 1.7.1

Therefore,

$$AB = \frac{\sigma(A)\sigma(B)}{\sqrt{\lambda_1\lambda_2}(\alpha-\beta)}\begin{pmatrix} \lambda_1(\lambda_2\alpha-\beta) & \lambda_1(1-\lambda_2)\alpha\beta \\ (\lambda_2-1) & (\alpha-\lambda_2\beta) \end{pmatrix}.$$

The roots of the characteristic equation of this matrix have the same sign coinciding with the sign of

$$\lambda_1(\lambda_2\alpha-\beta) + (\alpha-\lambda_2\beta) > 0.$$

Hence

$$\sigma(AB) = \sigma(A)\sigma(B). \qquad \square$$

THEOREM 7.1. *Suppose* $\{A_i, B_i \ (i=1,\dots,g), \ C_i \ (i=g+1,\dots,n)\} \subset$ $\mathrm{SL}(2,\mathbb{R})$ *and let* $\{J(A_i), J(B_i) \ (i=1,\dots,g), J(C_i) \ (i=g+1,\dots,n)\}$ *be a sequential set of type* $(g,k,m)$. *Then*

$$\prod_{i=1}^{g}[A_i,B_i] \prod_{i=g+1}^{n} C_i = 1$$

*if and only if either* $n=g$ *or*

$$\prod_{i=g+1}^{n} \sigma(C_i) = (-1)^{n-g}.$$

Proof. By Lemma 7.2, $\sigma([A_i, B_i]) = -1$. Therefore, it suffices to prove the theorem in the case $g=0$. In this case we proceed by induction on $n$. For $n=3$ the statement follows from Lemma 7.1. The induction step also follows from Lemma 7.1 applied to the set $\{C_1,\dots,C_{n-2},C_{n-1},C_n\}$. $\qquad \square$

**2.** Now we associate to a lifting of a Fuchsian group

$$\Gamma \subset \mathrm{Aut}_2(\Lambda) \cup \mathrm{Aut}_1(\Lambda)$$

a function

$$H_1(\Lambda/\Gamma, \mathbb{Z}_2) \to \mathbb{Z}_2,$$

where $\mathbb{Z}_2 = \mathbb{Z}/2\mathbb{Z} = \{0,1\}$. Let $P = \Lambda/\Gamma$ and let $\Gamma^* \subset \mathrm{SL}(2,\mathbb{R})$ be a subgroup such that $J$ induces an isomorphism $\tilde{J} \colon \Gamma^* \to \Gamma$. For a given point $q \in \Lambda$, an uniformization $\Phi \colon \Lambda \to P$ induces an isomorphism $\Phi_q \colon \Gamma \to \pi_1(P,p)$, where $p = \Phi(q)$ (see Section 2). Construct a function $\tilde{\omega} \colon \pi_1(P,p) \to \mathbb{Z}_2$ by setting

$$\tilde{\omega}(c) = \begin{cases} 0 & \text{if } \sigma(\tilde{J}^{-1}(\Phi_q^{-1}(c))) = -1, \\ 1 & \text{if } \sigma(\tilde{J}^{-1}(\Phi_q^{-1}(c))) = 1. \end{cases}$$

We denote by $(.,.) \colon H_1(P,\mathbb{Z}_2) \times H_2(P,\mathbb{Z}_2) \to \mathbb{Z}_2$ the intersection form in $H_1(P,\mathbb{Z}_2)$. The natural projection $\pi_1(P,p) \to H_1(P,\mathbb{Z}_2)$ takes $(.,.)$ to a pairing on $\pi_1(P,p)$.

LEMMA 7.3. *Let* $\tilde{c}_1, \tilde{c}_2, \tilde{c}_3 \subset P$ *be simple contours such that* $\tilde{c}_i \cap \tilde{c}_j = p$ *and for the corresponding elements* $c_i \in \pi_1(P,p)$ *we have* $c_i \neq 1$ *and* $c_1c_2c_3 = 1$. *Then* $\tilde{\omega}(c_3) = \tilde{\omega}(c_1) + \tilde{\omega}(c_2) + (c_1,c_2)$.

Proof. By Theorem 2.1, the shifts $\Phi_q^{-1}(c_i)$ form a sequential set either of type $(0, k, m)$ or of type $(2, 1, 0)$. In the first case $(c_1, c_2) = 0$ and $\tilde{\omega}(c_3) = \tilde{\omega}(c_1) + \tilde{\omega}(c_2)$, by Lemma 7.1. In the second case $(c_1, c_2) = 1$ and $\tilde{\omega}(c_3) = \tilde{\omega}(c_1) + \tilde{\omega}(c_2) + 1$, by Lemma 7.2. $\square$

LEMMA 7.4. *Suppose* $d_1, d_2 \in \pi_1(P, p)$ *represent the same element of* $H_1(P, \mathbb{Z}_2)$. *Then* $\tilde{\omega}(d_1) = \tilde{\omega}(d_2)$.

Proof. Consider a standard system of generators

$$v = \{a_i, b_i \ (i = 1, \ldots, g), \ c_i \ (i = g + 1, \ldots, n)\}$$

of $\pi_1(P, p)$. Consider a function $\omega_v \colon H_1(P, \mathbb{Z}_2) \to \mathbb{Z}_2$ coinciding with $\tilde{\omega}$ on $v$ and possessing the property $\omega_v(a + b) = \omega_v(a) + \omega_v(b) + (a, b)$ for any $a, b \in H_1(P, \mathbb{Z}_2)$. These properties determine $\omega_v$ uniquely. Consider Dehn's twists [17], that is, transformations of one of the following types:

1. $a_1 \mapsto a_1 b_1$.
2. $a_1 \mapsto a_1 a_2 a_1 a_2^{-1} a_1^{-1}$,
   $b_1 \mapsto a_1 a_2 a_1^{-1} a_2^{-1} b_1 a_2^{-1} a_1^{-1}$,
   $a_2 \mapsto a_1 a_2 a_1^{-1}$,
   $b_2 \mapsto b_2 a_2^{-1} a_1^{-1}$.
3. $a_g \mapsto b_g^{-1} c_1 b_g^{-1} c_1^{-1} b_g$,
   $b_g \mapsto b_g^{-1} c_1 b_g c_1^{-1} b_g a_g b_g^{-2} c_g^{-1} b_g$,
   $c_1 \mapsto b_g^{-1} c_1 b_g$.

Such a transformation takes $v$ to another standard system of generators $\tilde{v}$. Lemma 7.3 implies that $\omega_{\tilde{v}} = \omega_v$. On the other hand, using Dehn's twists and renumbering the generators, one can obtain from $v$ any standard system of generators, in particular, a system containing any prescribed simple contour that does not belong to the kernel of $(.\,,.)$ as $a_1$, and any prescribed simple contour that is homotopy equivalent to a hole or a puncture as $c_{g+1}$. Hence $\tilde{\omega} = \omega_v \eta$, where $\eta \colon \pi_1(P, p) \to H_1(P, \mathbb{Z}_2)$ is a natural projection. $\square$

By Lemma 7.4, the function $\tilde{\omega}$ induces a function $\omega \colon H_1(P, \mathbb{Z}_2) \to \mathbb{Z}_2$. Lemma 7.3 immediately implies

COROLLARY 7.1. $\omega(a + b) = \omega(a) + \omega(b) + (a, b)$.

Functions $\omega \colon H_1(P, \mathbb{Z}_2) \to \mathbb{Z}_2$ satisfying this condition are called *Arf functions* [23]. Hence, our construction associates to each lifting $\Gamma^*$ of $\Gamma$ an Arf function $\omega_{\Gamma^*} \colon H_1(\Lambda/\Gamma, \mathbb{Z}_2) \to \mathbb{Z}_2$.

THEOREM 7.2 ([70]). *The correspondence* $\Gamma^* \mapsto \omega_{\Gamma^*}$ *between liftings of a group* $\Gamma$ *and Arf functions on* $H_1(\Lambda/\Gamma, \mathbb{Z}_2)$ *is one-to-one.*

Proof. It is clear from the construction that distinct Arf functions correspond to distinct liftings. Now let us associate to an arbitrary Arf function $\omega$ a lifting $\Gamma^*$ such that $\omega_{\Gamma^*} = \omega$. To this end, consider an arbitrary sequential set

$$V = \{A_i, B_i \ (i = 1, \ldots, g), \ C_i \ (i = g + 1, \ldots, n)\} = \{D_1, \ldots, D_{g+n}\},$$

that generates $\Gamma$. Consider the matrices

$$V^* = \{D_1^*, \dots, D_{g+n}^*\} \in \mathrm{SL}(2, \mathbb{R})$$

which are uniquely defined by the conditions

$$J(D_i^*) = D_i \quad \text{and} \quad \sigma(D_i^*) = \begin{cases} -1 & \text{if } \omega(\Phi_q(D_i)) = 0, \\ 1 & \text{if } \omega(\Phi_q(D_i)) = 1. \end{cases}$$

By Lemmas 7.3, 7.1, we have

$$\prod_{i=1}^{g} [A_i^*, B_i^*] \prod_{i=g+1}^{n} C_i^* = 1.$$

Since

$$\prod_{i=1}^{g} [A_i, B_i] \prod_{i=g+1}^{n} C_i = 1$$

is the only generating relation in $\Gamma$, the restriction of $J$ to the group $\Gamma^*$ generated by $V^*$ induces an isomorphism $\Gamma^* \to \Gamma$. By construction, $\omega_{\Gamma^*} = \omega$.

$\square$

## 8. Topological classification
## of Arf functions and pairs of Arf functions

In the previous section we reduced the description of lifted Fuchsian groups to that of Arf functions, that is, mappings $\omega \colon H_1(P, \mathbb{Z}_2) \to \mathbb{Z}_2$ such that

$$\omega(a + b) = \omega(a) + \omega(b) + (a, b),$$

where $\mathbb{Z}_2 = \mathbb{Z}/2\mathbb{Z}$ and $(a, b) \in \mathbb{Z}_2$ is the intersection number of $a$ and $b$ in $H_1(P, \mathbb{Z}_2)$. Autohomeomorphisms $\phi \colon P \to P$ act on the set of Arf functions. The aim of the present section consists in describing the orbits of this action.

Let $P$ be a surface of type $(g, k, m)$. We say that a basis

$$v = \{a_i, b_i \ (i = 1, \dots, g), c_i \ (i = g+1, \dots, n)\}$$

of $H_1(P, \mathbb{Z}_2)$ is *standard* if it is the image of a standard system of generators

$$v' = \{a_i', b_i' \ (i = 1, \dots, g), c_i' \ (i = g+1, \dots, n)\} \in \pi_1(P, p)$$

under the natural projection $\pi_1(P, p) \to H_1(P, \mathbb{Z}_2)$.

Consider the group $\mathrm{Mod}(P)$ of autohomeomorphisms of $P$ taking holes to holes and punctures to punctures. It determines a group $A(P)$ which acts faithfully on the set of Arf functions. An element of $A(P)$ is uniquely determined by its action on a standard basis $v$. The Dehn generators [17] of $\mathrm{Mod}(P)$ determine generators of $A(P)$. They split into five types and are uniquely determined by the image

$$\tilde{v} = \{\tilde{a}_i, \tilde{b}_i \ (i = 1, \dots, g), \tilde{c}_i \ (i = g+1, \dots, n)\} \subset H_1(P, \mathbb{Z}_2)$$

of a given standard basis

$$v = \{a_i, b_i \ (i = 1, \dots, g), c_i \ (i = g+1, \dots, n)\} \subset H_1(P, \mathbb{Z}_2).$$

Here are the five types of transformations:

(1) $\tilde{a}_i = a_i + b_i$; (2) $\tilde{a}_i = b_i$, $\tilde{b}_i = a_i$;

(3) $\tilde{a}_i = a_i + a_j$, $\tilde{b}_j = b_j$, $\tilde{a}_i = a_i, \tilde{b}_i = b_i + b_j$;

(4) $\tilde{a}_i = a_i + c_r$; (5) $\tilde{c}_i = c_j$, $\tilde{c}_j = c_i$,

where $\tilde{a}_k = a_k$, $\tilde{b}_k = b_k$, $\tilde{c}_k = c_k$ for $k \neq i, j$.

Let us set $\delta = \delta(P, \omega) = 0$ if there is a standard basis $v$ such that

$$\sum_{i=1}^{g} \omega(a_i)\omega(b_i) \equiv 0 \pmod{2}.$$

In other cases we set $\delta = 1$.

The set of holes $c_{g+1}, \ldots, c_{g+k}$ decomposes into two subsets,

$$D_0 = \{c_i \mid \omega(c_i) = 0\}$$

and

$$D_1 = \{c_i \mid \omega(c_i) = 1\}.$$

Denote by $k_\alpha = k_\alpha(P, \omega)$ the cardinality of $D_\alpha$. Similarly, the punctures $c_{g+k+1}, \ldots, c_n$ are separated into two subsets of the form

$$E_\alpha = \{c_i \mid \omega(c_i) = \alpha\}.$$

Let $m_\alpha$ ($\alpha = 0, 1$) denote the cardinality of $E_\alpha$. The *type of an Arf function* $(P, \omega)$ is the set $(g, \delta, k_0, k_1, m_0, m_1)$. Set $n_\alpha = k_\alpha + m_\alpha$.

LEMMA 8.1. *A set $(g, \delta, k_\alpha, m_\alpha)$ is a type of an Arf function if and only if*

(1) $k_1 + m_1 \equiv 0 \pmod{2}$;

(2) $\delta = 0$ *for* $n_1 > 0$.

*If these conditions are satisfied, then there is a standard basis $v$ such that* $\omega(a_i) = \omega(b_i) = 0$ *for* $i > 1$, $\omega(a_1) = \omega(b_1) = \delta$, $\omega(c_i) = 0$ *for* $g + 1 \leqslant i \leqslant g + k_0$, *and* $g + k_0 + k_1 + 1 \leqslant i \leqslant g + k_0 + k_1 + m_1$.

Proof. The condition $k_1 + m_1 \equiv 0 \pmod{2}$ follows from the equations

$$\sum_{i=g+1}^{n} \omega(c_i) = \omega\left(\sum_{i=g+1}^{n} c_i\right) = \omega(0) = 0.$$

Now let us prove that there is a standard basis $v$ such that $\omega(a_i) = \omega(b_i) = 0$ for $i > 1$ and $\omega(a_1) = \omega(b_1) = 0$ for $n_1 > 0$. It is easy to see that, by using transformations (1)–(3), we can turn an arbitrary standard basis $v$ into a standard basis such that $\omega(a_i) = \omega(b_i) = 0$ for $i > 1$ and $\omega(a_1) = \omega(b_1)$. If, moreover, $\omega(a_1) = \omega(b_1) = \omega(c_i) = 1$, then the transformation $\tilde{a}_1 = a_1 + c_i$, $\tilde{b}_1 = b_1 + c_i$ leads to a desired basis. If $n_1 = 0$, then the transformations (1)–(5) preserve the parity of the sum

$$\sum_{i=1}^{g} \omega(a_i)\omega(b_i),$$

whence the type $(g, 1, k_0, 0, m_0, 0)$ is realizable.                                      $\square$

Now let $(P, \omega_1, \omega_2)$ be a pair of Arf functions, and suppose $\omega_1 \neq \omega_2$. The *type* of the pair is the set $(g, \delta_1, \delta_2, k_{\alpha\beta}, m_{\alpha\beta})$, where $\delta_i = \delta(P, \omega_i)$, $k_{\alpha\beta}$ is the number of holes $c_i$ such that $\omega_1(c_i) = \alpha$, $\omega_2(c_i) = \beta$, and $m_{\alpha\beta}$ is the number of punctures $c_i$ such that $\omega_1(c_i) = \alpha$, $\omega_2(c_i) = \beta$, where $\alpha, \beta \in \{0, 1\}$. We set $n_{\alpha\beta} = k_{\alpha\beta} + m_{\alpha\beta}$.

LEMMA 8.2. *A set* $(g, \delta_1, \delta_2, k_{\alpha\beta}, m_{\alpha\beta})$ *is a type of a pair of Arf functions if and only if* $(g, \delta_1, k_{\alpha 0} + k_{\alpha 1}, m_{\alpha 0} + m_{\alpha 1})$ *and* $(g, \delta_2, k_{0\alpha} + k_{1\alpha}, m_{0\alpha} + m_{1\alpha})$ *are types of some Arf functions. If this condition is satisfied, then there is a standard basis* $v$ *such that* : (1) *if* $n_{10} + n_{01} > 0$, *then* $\omega_j(a_1) = \omega_j(b_1) = \delta_j$, $\omega_j(a_i) = \omega_j(b_i) = 0$, *where* $j = 1, 2$, $i > 1$; (2) *if* $n_{10} = n_{01} = 0$ *and* $n_{11} > 0$, *then* $\omega_j(a_i) = \omega_j(b_i) = 0$ *for* $j = 1, 2$, $i > 1$ *and* $\omega_1(b_1) = \omega_2(a_1) = \omega_2(b_1) = 0$, $\omega_1(a_1) = 1$; (3) *if* $n_{01} = n_{10} = n_{11} = 0$, *then* $\omega_1(a_1) + \omega_2(a_1) = 1$, $\omega_1(a_2) = \omega_1(b_2) = \omega_2(a_2) = \omega_2(b_2) = \varepsilon$, $\omega_1(b_1) = \omega_2(b_1) = \delta$ (*where* $\varepsilon = 0$ *for* $\delta = 1$, *and* $\omega_1(a_1) = 1$ *for* $\delta = 0$) *and* $\omega_k(a_i) = \omega_k(b_i) = 0$ *for* $i > 2$.

Proof. The first statement is obvious. Let $(P, \omega_1, \omega_2)$ be a pair of distinct Arf functions. Consider the epimorphism $\omega = (\omega_1 + \omega_2) \colon H_1(P, \mathbb{Z}_2) \to \mathbb{Z}_2$. It is easy to transform, using (1)–(5), an arbitrary basis into a basis

$$v = \{a_i, b_i, c_j\} \subset H_1(P, \mathbb{Z}_2)$$

such that

$$\omega(b_1) = \omega(a_i) = \omega(b_i) = 0 \quad \text{for} \quad i > 1$$

and

$$\omega(a_1) = 0 \quad \text{for} \quad n_{01} + n_{10} > 0.$$

Suppose $n_{01} + n_{10} > 0$. Then $\omega_1(a_i) = \omega_2(a_i)$ and $\omega_1(b_i) = \omega_2(b_i)$. Transformations (1)–(3) lead to a basis $v$ such that

$$\omega_k(a_i) = \omega_k(b_i) = 0 \quad \text{for} \quad i > 1$$

and

$$\omega_1(a_1) = \omega_1(b_1) = \omega_2(a_1) = \omega_2(b_1).$$

Using, if necessary, the transformations $\tilde{a}_1 = a_1 + c_k$ and $\tilde{b}_1 = b_1 + c_k$, we obtain $\omega_k(\tilde{a}_1) = \omega_k(\tilde{b}_1) = \delta_k$. Now suppose $n_{01} = n_{10} = 0$ and let $b \subset P$ be a simple contour representing the element $b_1$. Then the functions $\omega_1$ and $\omega_2$ coincide on $P \setminus b$ and, by Lemma 8.1, there is a basis $v \subset H_1(P, \mathbb{Z}_2)$ such that (1) $\omega_k(a_i) = \omega_k(b_i) = 0$ for $i > 2$; (2) $\omega_1(a_2) = \omega_1(b_2) = \omega_2(a_2) = \omega_2(b_2) = \varepsilon$, $\omega_1(b_1) = \omega_2(b_1) = \delta$, $\omega_1(a_1) + \omega_2(a_1) = 1$ and $\varepsilon = 0$ if either $\delta = 1$ or $n_{11} > 0$. If $n_{11} > 0$, then, for an appropriate $k$, the transformation $\tilde{b}_1 = b_1 + c_k$ yields $\delta = 0$. If $\delta = 0$, then the transformation $\tilde{a}_1 = a_1 + b_1$ yields $\omega_1(\tilde{a}_1) = 1$.  □

Two tuples of Arf functions $(\omega_1, \ldots, \omega_n)$ and $(\omega'_1, \ldots, \omega'_n)$ on $P$ are said to be *topologically equivalent* if there is a homeomorphism $\alpha \in \mathrm{Mod}(P)$ such that $\omega'_i = \omega_i \alpha$.

THEOREM 8.1 ([**76**], [**74**]). *Arf functions and pairs of Arf functions are equivalent if and only if they have the same topological type.*

Proof. Transformations (1)–(5) preserve the topological type of Arf functions and pairs of Arf functions, which can be checked directly. These transformations generate the group $A(P)$. Therefore, the topological type is preserved under the action of any $\alpha \in \mathrm{Mod}\ (P)$. Now suppose $\omega$ and $\omega'$ have the same topological type. Then, by Lemma 8.1, there are standard bases $v = \{v_i\}$ and $v' = \{v_i'\}$ of $\pi_1(P, p)$ such that $\omega(v_i) = \omega'(v_i')$. Consider an element $\alpha \in \mathrm{Mod}(P)$ such that $\alpha(v_i') = v_i$. Then $\omega' = \omega\alpha$ and, therefore, $\omega'$ and $\omega$ are topologically equivalent. The second statement of Theorem 8.1 is proved in a similar way on the base of Lemma 8.2.    □

The case of compact surfaces is the most important one, and we restrict ourselves to this case till the end of the present section. According to our definitions, in this case the topological type of an Arf function is described by an element $\delta(\omega) \in \{0, 1\}$, called the *Arf invariant* [2].

LEMMA 8.3. *Let $\{a_i, b_i\ (i = 1, \ldots, g)\}$ be an arbitrary standard basis in $H_1(P, \mathbb{Z}_2)$. Then*

$$\delta(\omega) \equiv \sum_{i=1}^{g} \omega(a_i)\omega(b_i) \quad (\mathrm{mod}\ 2).$$

Proof. The right-hand side of the above congruence is invariant under transformations (1)–(5). On the other hand, these transformations generate a group which acts transitively on the set of all standard bases.    □

For the case of compact surfaces, Theorem 8.1 and Lemma 8.2 yield

COROLLARY 8.1. *Two Arf functions on a compact surface are topologically equivalent if and only if their Arf invariants coincide. Two pairs of Arf functions on a compact surface are topologically equivalent if and only if their Arf invariants coincide.*

In the next section we shall prove the analog of this statement for the case of an arbitrary number of Arf functions.

## 9. Topological classification of independent Arf functions on compact surfaces

In this section, the words "an Arf function" refer to an Arf function on a given compact Riemann surface $P$ of genus $g$. The sum of evenly many such functions is a linear form $H_1(P, \mathbb{Z}_2) \to \mathbb{Z}_2$. The sum of oddly many Arf functions is an Arf function itself.

A set
$$\Omega = (\omega_0, \omega_1, \ldots, \omega_n)$$
of Arf functions is said to be *independent* if the identity

$$\sum_{i=0}^{n} \alpha_i \omega_i = 0,$$

where $\alpha_i \in \{0, 1\}$, implies $\alpha_0 = \alpha_1 = \cdots = \alpha_n = 0$. Our aim is to prove the following theorem.

THEOREM 9.1 ([**77**]). *Two independent sets of Arf functions*

$$\Omega^1 = (\omega_0^1, \ldots, \omega_n^1) \quad \text{and} \quad \Omega^2 = (\omega_0^2, \ldots, \omega_n^2)$$

*are topologically equivalent if and only if*

$$\delta(\omega_0^1 + \omega_i^1 + \omega_j^1) = \delta(\omega_0^2 + \omega_i^2 + \omega_j^2) \quad \text{for all} \quad i, j = 0, 1, \ldots, n.$$

Indeed, consider $H_1(P, \mathbb{Z}_2)$ as a vector space $H$ over $\mathbb{Z}_2$. The intersection number induces a nondegenerate symmetric bilinear form $(.,.): H \times H \to \mathbb{Z}_2$ on $H$ such that $(a, a) = 0$. A basis

$$\{a_i, b_i \ (i = 1, \ldots, g)\} \subset H$$

is said to by *symplectic* if $(a_i, a_j) = (b_i, b_j) = 0$, $(a_i, b_j) = \delta_{ij}$.

Let $V = (v_1, \ldots, v_n) \subset H$ be a set of linearly independent vectors. Here is the "diagonalization procedure" for such a set. The result of the procedure is a set $W = W(V) = \{s_i, t_i \ (i = 1, \ldots, r), s_i \ (i = r + 1, \ldots, r + h)\}$ constructed inductively starting from $V^0 = V$.

The passage from

$$V^{k-1} = \{s_i, t_i \ (i = 1, \ldots, k - 1), v_{j_1}^{k-1}, \ldots, v_{j_{n_{k-1}}}^{k-1}\}$$

to

$$V^k = \{s_i, t_i \ (i = 1, \ldots, k), v_{j_1}^k, \ldots, v_{j_{n_k}}^k\}$$

is described by the following rules.

If all $v_i^{k-1}$ are pairwise orthogonal (i.e., $(v_i^{k-1}, v_j^{k-1}) = 0$), then we set

$$W = V^{k-1}.$$

Otherwise we proceed as follows:

(1) choose the vector $v_m^{k-1}$ with the minimal subscript $m = i$, which is not orthogonal to at least one of the $v_m^{k-1}$;

(2) choose the vector $v_m^{k-1}$ with the minimal subscript $m = j$, such that $(v_i^{k-1}, v_j^{k-1}) = 1$;

(3) set $s_k = v_i^{k-1}$, $t_k = v_j^{k-1}$;

(4) for $m \neq i, j$ set $v_m^k = v_m^{k-1} - (v_m^{k-1}, s_k)t_k + (v_m^{k-1}, t_k)s_k$.

LEMMA 9.1. *Suppose two sets of linearly independent vectors*

$$V = (v_1, \ldots, v_n) \quad \text{and} \quad V' = (v_1', \ldots, v_n')$$

*are such that* $(v_i, v_j) = (v_i', v_j')$ *for all* $i, j$. *Let* $\omega$ *and* $\omega'$ *be two Arf functions such that* $\delta(\omega) = \delta(\omega')$ *and* $\omega(v_i) = \omega'(v_i')$ *for* $i = 1, \ldots, n$. *Then there is an isomorphism* $\phi: H \to H$ *such that* $(a, b) = (\phi(a), \phi(b))$, $\omega'\phi = \omega$ *and* $\phi(v_i) = \phi(v_i')$ *for all* $a, b \in H$, $v_i \subset V$.

Proof. Set

$$W(V) = \{s_i, t_i \ (i = 1, \ldots, r), s_i \ (i = r + 1, \ldots, r + h)\}$$

and

$$W(V') = \{s_i', t_i' \ (i = 1, \ldots, r), s_i' \ (i = r + 1, \ldots, r + h)\}.$$

Then the equation $\omega(v_i) = \omega'(v_i')$ implies $\omega(s_i) = \omega'(s_i')$, $\omega(t_i) = \omega'(t_i')$. Besides, $(s_i, s_j) = (t_i, t_j) = (s_i', s_j') = (t_i', t_j') = 0$, $(s_i, t_j) = (s_i', t_j') = \delta_{ij}$, and so $W(V)$ and $W(V')$ can be extended to symplectic bases
$$\{a_i, b_i \ (i = 1, \ldots, g)\}, \quad \{a_i', b_i' \ (i = 1, \ldots, g)\}.$$
Using the condition $\delta(\omega) = \delta(\omega')$ and repeating the argument in the proof of Lemma 8.1, it is easy to show that the complementary vectors can be chosen in such a way that $\omega(a_i) = \omega'(a_i')$ and $\omega(b_i) = \omega'(b_i')$ for all $i$. The correspondence $a_i \mapsto a_i'$, $b_i \mapsto b_i'$ generates an isomorphism $\phi \colon H \to H$ with the desired properties. $\qquad \square$

We mark by $*$ the objects dual to given ones with respect to the bilinear form $(.\,,.)$. In particular, if $\omega'$ and $\omega''$ are Arf functions, then $\omega' + \omega'' \in H^*$ and $(\omega' + \omega'')^* \in H$.

LEMMA 9.2. *Let* $\{\omega_0, \omega_1, \omega_2\}$ *be Arf functions. Then*
(1) $\omega_0((\omega_0 + \omega_1)^*) = \delta(\omega_0) + \delta(\omega_1)$;
(2) $\delta(\omega_0 + \omega_1 + \omega_2) = \delta(\omega_0) + \delta(\omega_1) + \delta(\omega_2) + ((\omega_0 + \omega_1)^*, (\omega_0 + \omega_2)^*)$.

Proof. Let $\{a_i, \ b_i \ (i = 1, \ldots, g)\}$ be a symplectic basis in $H$ such that $a_1 = (\omega_0 + \omega_1)^*$. Then $\omega_1 = \omega_0 + a_1^*$, and therefore, $\omega_1(a_i) = \omega_0(a_i)$ for all $i$, $\omega_1(b_i) = \omega_0(b_i)$ for $i > 1$, and $\omega_1(b_1) = \omega_0(b_1) + 1$. Thus,
$$\delta(\omega_0) + \delta(\omega_1) = \omega_0(a_1)\omega_0(b_1) + \omega_1(a_1)\omega_1(b_1) = \omega_0(a_1) = \omega_0((\omega_0 + \omega_1)^*).$$
Hence,
$$\begin{aligned}
\delta(\omega_0 + \omega_1 + \omega_2) &+ \delta(\omega_0) + \delta(\omega_1) + \delta(\omega_2) \\
&= (\delta(\omega_0) + \delta(\omega_0 + \omega_1 + \omega_2)) + (\delta(\omega_0) + \delta(\omega_1)) + (\delta(\omega_0) + \delta(\omega_2)) \\
&= \omega_0((\omega_1 + \omega_2)^*) + \omega_0((\omega_0 + \omega_1)^*) + \omega_0((\omega_0 + \omega_2)^*) \\
&= \omega_0((\omega_0 + \omega_1)^* + (\omega_0 + \omega_2)^*) + \omega_0((\omega_0 + \omega_1)^*) + \omega_0((\omega_0 + \omega_2)^*) \\
&= ((\omega_0 + \omega_1)^*, (\omega_0 + \omega_2)^*).
\end{aligned}$$

$\qquad \square$

Proof of Theorem 9.1. In one direction, the statement is obvious. Now suppose the sets $\Omega^1$ and $\Omega^2$ are such that
$$\delta(\omega_0^1 + \omega_i^1 + \omega_j^1) = \delta(\omega_0^2 + \omega_1^2 + \omega_2^2)$$
and, in particular, $\delta(\omega_i^1) = \delta(\omega_i^2)$ for all $i, j = 0, 1, \ldots, n$.
   We set
$$v_i^j = (\omega_0^j + \omega_i^j)^* \quad \text{and} \quad V^j = (v_1^j, \ldots, v_n^j).$$
Then, by Lemma 9.2, we have
$$(v_i^1, v_j^1) = (v_i^2, v_j^2) \quad \text{and} \quad \omega_0^1(v_i^1) = \omega_0^2(v_i^2).$$
Hence, according to Lemma 9.1, there is an isomorphism $\phi \colon H \to H$ preserving the bilinear form $(.\,,.)$, taking $V^1$ to $V^2$ and such that $\omega_0^2\phi = \omega_0^1$. Using the Dehn twists [17], it is easy to show that any automorphism $\phi$ preserving $(.\,,.)$ is generated by a homeomorphism $\tilde{\phi} \colon P \to P$. $\qquad \square$

## 10. The moduli space of spinor framings

**1.**    Let $e: E \to P$ be a locally trivial line bundle over a Riemann surface $P$. The uniformization $\Lambda \to \Lambda/\Gamma$, where $\Gamma \subset \mathrm{PSL}(2, \mathbb{R})$ is a Fuchsian group, lifts $e$ to a bundle $\tilde{e}: \tilde{E} \to \Lambda$. The latter admits a trivialization, i.e., there is a biholomorphic mapping $\tilde{\Phi}: \tilde{E} \to \Lambda \times \mathbb{C}$ taking $\tilde{e}$ to the natural projection $\tilde{\lambda}: (\Lambda \times \mathbb{C}) \to \Lambda$. Hence $e$ is isomorphic to a bundle that can be obtained by factorization of the trivial bundle $(\Lambda \times \mathbb{C})$ modulo an action of a group $\Gamma$, where an element $\gamma \in \Gamma \subset \mathrm{PSL}(2, \mathbb{R})$ acts on $(\Lambda \times \mathbb{C})$ according to the rule

$$\gamma(z, x) = \left( \frac{az + b}{cz + d}, f(\gamma, z) \cdot x \right).$$

The function $f: \Gamma \times \Lambda \to C^* = C \setminus \{0\}$ is called the *transition function*.

In particular, if $E$ is the cotangent bundle, then it is possible to choose $\tilde{\lambda}$ so that to make

$$f(\gamma, z) = \left( \frac{d\gamma}{dz} \right)^{-1} = (cz + d)^2.$$

If the functions $f$ and $g$ correspond to two mappings $E \to P$ and $F \to P$, then the transition function corresponding to the tensor product $E \otimes F \to P$ is $f \cdot g$.

A line bundle $E \to P$ is called a *spinor bundle* if its tensor square $E \otimes E \to P$ is isomorphic to the cotangent bundle. Thus, the square of a transition function for a spinor bundle has the form $f(\gamma, z) = \sigma(a, b, c, d)(cz + d)$, where $\sigma^2(a, b, c, d) = 1$.

We call a line bundle with a chosen 'uniformization' like above a *framing*. Hence a framing is determined by a transition function $f: \Gamma \times \Lambda \to C^*$. A *spinor framing* is a framing $f(\gamma, z) = \sigma(a, b, c, d)(cz + d)$, where $\sigma^2(a, b, c, d) = 1$. Associate to a mapping $\gamma(z) = \frac{az+b}{cz+d}$ in $\Gamma$ the matrix

$$J_f^*(\gamma) = \sigma(a, b, c, d) \cdot \begin{pmatrix} a & b \\ c & d \end{pmatrix} \in \mathrm{SL}(2, \mathbb{R}).$$

LEMMA 10.1. *The mapping* $J_f^*: \Gamma \to \mathrm{SL}(2, \mathbb{R})$ *is well defined and it is a monomorphism.*

Proof. To prove that $J^*$ is well defined means to show that

$$\sigma(-a, -b, -c, -d) = -\sigma(a, b, c, d),$$

which follows immediately from the definitions. If

$$\gamma_1(z) = \frac{a_1 z + b_1}{c_1 z + d_1}, \quad \gamma_2(z) = \frac{a_2 z + b_2}{c_2 z + d_2},$$

then

$$\gamma_1 \gamma_2(z) = \frac{\tilde{a}z + \tilde{b}}{\tilde{c}z + \tilde{d}}, \quad \text{where} \quad \begin{pmatrix} \tilde{a} & \tilde{b} \\ \tilde{c} & \tilde{d} \end{pmatrix} = \begin{pmatrix} a_1 & b_1 \\ c_1 & d_2 \end{pmatrix} \begin{pmatrix} a_2 & b_2 \\ c_2 & d_2 \end{pmatrix}.$$

Therefore,

$$\gamma_1 \gamma_2(z, x) = \left( \frac{\tilde{a}z + \tilde{b}}{\tilde{c}z + \tilde{d}}, f(\gamma_1 \gamma_2, z)x \right) = \left( \frac{\tilde{a}z + \tilde{b}}{\tilde{c}z + \tilde{d}}, \sigma(\tilde{a}, \tilde{b}, \tilde{c}, \tilde{d}) (\tilde{c}z + \tilde{d}) \right).$$

On the other hand,

$$\gamma_1\gamma_2(z,x) = \gamma_1\Big(\frac{a_2z+b_2}{c_2z+d_2},\ f(\gamma_2,z)x\Big)$$

$$= \Big(\frac{\tilde{a}z+\tilde{b}}{\tilde{c}z+\tilde{d}},\ \sigma(a_1,b_1,c_1,d_1)\Big(c_1\frac{a_2z+b_2}{c_2z+d_2}+d_1\Big)\sigma(a_2,b_2,c_2,d_2)(c_2z+d_2)x\Big).$$

Hence

$$\sigma(\tilde{a},\tilde{b},\tilde{c},\tilde{d})(\tilde{c}z+\tilde{d}) = \sigma(a_1,b_1,c_1,d_1)\sigma(a_2,b_2,c_2,d_2)(\tilde{c}z+\tilde{d})$$

and

$$\sigma(\tilde{a},\tilde{b},\tilde{c},\tilde{d})\begin{pmatrix}\tilde{a}&\tilde{b}\\\tilde{c}&\tilde{d}\end{pmatrix} = \sigma(a_1,b_1,c_1,d_1)\sigma(a_2,b_2,c_2,d_2)\begin{pmatrix}a_1&b_1\\c_1&d_1\end{pmatrix}\begin{pmatrix}a_2&b_2\\c_2&d_2\end{pmatrix}.$$

Thus, $J_f^*$ is a homomorphism. If $J\colon \mathrm{SL}(2,\mathbb{R}) \to \mathrm{PSL}(2,\mathbb{R})$ is the natural projection, then $JJ_f^* = \mathrm{id}$, whence $J_f^*$ is a monomorphism. $\square$

By Lemma 10.1, $J_f^*(\Gamma)$ is a lifting of $\Gamma$. Consider the Arf function $\omega_f\colon H_1(\Lambda/\Gamma,\mathbb{Z}_2) \to \mathbb{Z}_2$ corresponding to this lifting (see Section 7).

THEOREM 10.1 ([**70**], [**74**]). *The mapping $f \mapsto \omega_f$ establishes a one-to-one correspondence between spinor framings on one hand and Arf functions on the Riemann surface $P = \Lambda/\Gamma$ on the other hand.*

Proof. Theorem 7.2 establishes a one-to-one correspondence between Arf functions on $P$ and liftings of the group $\Gamma$. Let us prove that the mapping $f \mapsto J_f^*(\Gamma)$ establishes a one-to-one correspondence between spinor framings and liftings of $\Gamma$. In order to do this, it suffices to associate to each lifting $J^*\colon \Gamma \to \mathrm{SL}(2,\mathbb{R})$ a spinor framing $f$ such that $J_f^* = J^*$. Such a framing is determined by the transition function $f(\gamma,z) = (cz+d)$, where $J^*(\gamma) = \begin{pmatrix}a&b\\c&d\end{pmatrix}$. $\square$

**Remark.** For compact surfaces, spinor framings are in one-to-one correspondence with spinor bundles. Another description of the correspondence $f \mapsto \omega_f$ for the compact case is contained in [**4**], [**35**], [**51**].

**2.** Now consider the set $Sp$ of all pairs $(\Lambda/\Gamma, f)$, where $f\colon \Gamma \times \Lambda \to \mathbb{C}^*$ is a spinor framing. We identify $(\Lambda/\Gamma', f') = (\Lambda/\Gamma, f)$ if there is an automorphism $h \in \mathrm{Aut}(\Lambda)$ such that $\Gamma' = h\Gamma h^{-1}$ and $f'(h\gamma h^{-1}, hz) = f(\gamma, z)$. By Theorem 10.1, a framing $(\Lambda/\Gamma, f)$ is determined by a pair $(P,\omega)$, where $\omega$ is an Arf function on $P = \Lambda/\Gamma$. Hence, the topology of the moduli space of Riemann surfaces induces a topology on $Sp$. Suppose a set $(g,\delta,k_\alpha,m_\alpha)$ satisfies all the assumptions of Lemma 8.1. Denote by $S(g,\delta,k_\alpha,m_\alpha)$ the set of all spinor framings $(\Lambda/\Gamma, f)$ such that $\omega_f$ is an Arf function of topological type $t = (g,\delta,k_\alpha,m_\alpha)$.

THEOREM 10.2 ([**74**]). *The space $S(g,\delta,k_\alpha,m_\alpha)$ is homeomorphic to $T_t/\mathrm{Mod}_t$, where*

$$T_t \cong T_{g,k_1+k_2,m_1+m_2} \cong \mathbb{R}^{6g+3(k_1+k_2)+2(m_1+m_2)-6}$$

*and*
$$\mathrm{Mod}_t \subset \mathrm{Mod}_{g,k_1+k_2,m_1+m_2}$$
*is a subgroup of finite index.*

Proof. Set $k = k_1 + k_2$, $m = m_1 + m_2$, $n = k + m + g$ and let $\psi \in T_{g,k,m}$. By definition, this means that $\psi \in \mathrm{Hom}(\gamma, \mathrm{Aut}(\Lambda))$, where the group $\gamma = \gamma_{g,n}$ is generated by $v_{g,n} = \{a_i, b_i \ (i = 1, \ldots, g), c_i \ (i = g+1, \ldots, n)\}$. Let
$$\tilde{v}_{g,n} = \{\tilde{a}_i, \tilde{b}_i \ (i = 1, \ldots, g), \tilde{c}_i \ (i = g+1, \ldots, n)\}$$
be the corresponding basis in $H_1(P, \mathbb{Z}_2)$, where $P = P_\psi = \Lambda/\psi(\gamma)$. Consider the Arf function $\omega = \omega_\psi$ on $P$ such that $\omega(a_i) = \omega(b_i) = 0$ for $i < g$, $\omega(a_g) = \omega(b_g) = \delta$, $\omega(c_j) = 0$ for $g + 1 \leqslant j \leqslant g + k_0$, $g + k + 1 \leqslant j \leqslant g + k + m_1$ and $\omega(c_j) = 1$ otherwise. By Theorem 10.1, a spinor framing $f_\psi \colon E_\psi \to P_\psi$ is associated to the Arf function $\omega_\psi$. The correspondence $\psi \mapsto f_\psi$ determines a mapping
$$\Psi \colon T_{g,k,m} \to S(g, \delta, k_\alpha, m_\alpha).$$
By Lemma 8.1,
$$\Psi(T_{g,k,m}) = S(g, \delta, k_\alpha, m_\alpha).$$
The preimage of each point $s \in S(g, \delta, k_\alpha, m_\alpha)$ is an orbit of the group $\mathrm{Aut}(P_\psi) = \mathrm{Mod}_{g,k,m}$ which preserves $\omega_\psi$. Hence
$$S(g, \delta, k_\alpha, m_\alpha) = T_{g,k,m}/\mathrm{Mod}_t,$$
where $\mathrm{Mod}_t$ is a discrete group. $\qquad\square$

## 11. Super-Fuchsian groups, super-Riemann surfaces and their topological types

Let $L = L(\mathbb{K})$ be the Grassmann algebra over a field $\mathbb{K}$, with infinitely many generators $1, \ell_1, \ell_2, \ldots$ . Each element $a \in L(\mathbb{K})$ is a finite linear combination of monomials $\ell_{i_1} \wedge \cdots \wedge \ell_{i_n}$ with coefficients in $\mathbb{K}$, i.e.,
$$a = a^\sharp + \sum a_i \ell_i + \sum_{ij} a_{ij} \ell_i \wedge \ell_j + \cdots .$$

The correspondence $a \mapsto a^\sharp$ determines an epimorphism $\sharp \colon L(\mathbb{K}) \to \mathbb{K}$.

A monomial $\ell_{i_1} \wedge \cdots \wedge \ell_{i_n} \neq 0$ is said to be *even* (respectively, *odd*) if $n$ is even (respectively, odd). Constants are treated as even monomials. Even (respectively, odd) monomials span the subspace $L_0(\mathbb{K})$ (respectively, $L_1(\mathbb{K})$) of even (respectively, odd) elements of $L(\mathbb{K})$. The superanalog of a vector space is the set
$$\mathbb{K}^{(n|m)} = \{(z_1, \ldots, z_n | \theta_1, \ldots, \theta_m) \colon z_i \in L_0(\mathbb{K}), \theta_j \in L_1(\mathbb{K})\}.$$

In what follows we assume that $\mathbb{K}$ is either the field of complex numbers $\mathbb{C}$ or the field of real numbers $\mathbb{R}$.

The set
$$\Lambda^{NS} = \{(z | \theta_1, \ldots, \theta_N) \in \mathbb{C}^{(1|N)} \,|\, \mathrm{Im}\ z^\sharp > 0\}$$

is called the *upper N-super half-plane*. In this section we shall discuss the 1-super half-plane $\Lambda^S = \Lambda^{1S}$. Its automorphism group $\operatorname{Aut}(\Lambda^S)$ consists of transformations $A = A[a, b, c, d, \sigma \,|\, \varepsilon, \delta]$ of the form

$$A(z \,|\, \theta) = \left( \frac{az + b}{cz + d} - \frac{(ad - bc)(\varepsilon + \delta z)}{(cz + d)^2} \theta \;\Big|\; \frac{\sigma \sqrt{ad - bc}}{cz + d} \left( \theta + \varepsilon + \delta z + \frac{1}{2} \varepsilon \delta \theta \right) \right),$$

where $a, b, c, d \in L_0(\mathbb{R})$, $\sigma = \pm 1$, $\varepsilon, \delta \in L_1(\mathbb{R})$, $(ad - bc)^\sharp > 0$, and $\sqrt{\Delta}$ denotes the element in $L_0(\mathbb{R})$ uniquely determined by the properties $(\sqrt{\Delta})^2 = \Delta$ and $(\sqrt{\Delta})^\sharp > 0$ (see [6], [7]).

The correspondence

$$A \mapsto A^\sharp, \quad \text{where} \quad A^\sharp(z) = \frac{a^\sharp z + b^\sharp}{c^\sharp z + d^\sharp},$$

induces the epimorphism

$$\sharp \colon \operatorname{Aut}(\Lambda^S) \to \operatorname{Aut}(\Lambda).$$

The elements of the set

$$\operatorname{Aut}_1(\Lambda^S) = (\sharp)^{-1}(\operatorname{Aut}_1(\Lambda))$$

are said to be *superparabolic*, while those of the set

$$\operatorname{Aut}_2(\Lambda^S) = (\sharp)^{-1}(\operatorname{Aut}_2(\Lambda))$$

are said to be *superhyperbolic*.

We say that a subgroup $\Gamma \subset \operatorname{Aut}(\Lambda^S)$ is *super-Fuchsian* if $\Gamma^\sharp = \sharp(\Gamma)$ is a Fuchsian group and $\sharp \colon \Gamma \to \Gamma^\sharp$ is an isomorphism. Later in this section we consider only super-Fuchsian groups consisting of superparabolic and superhyperbolic automorphisms of $\Lambda^S$.

Associate to an automorphism $A = A[a, b, c, d, \sigma \,|\, \varepsilon, \delta]$ the matrix

$$\bar{J}(A) = \frac{\sigma}{\sqrt{a^\sharp d^\sharp - c^\sharp d^\sharp}} \begin{pmatrix} a^\sharp & b^\sharp \\ c^\sharp & d^\sharp \end{pmatrix} \in \operatorname{SL}(2, \mathbb{R}).$$

LEMMA 11.1. *Let $\Gamma$ be a super-Fuchsian group. Then the correspondence $\bar{J} \colon \Gamma \to \operatorname{SL}(2, \mathbb{R})$ is a monomorphism, and therefore, it determines a lifting $J^* \colon \Gamma^\sharp \to \bar{\Gamma} = \bar{J}(\Gamma)$.*

Proof. A direct calculation shows that $\bar{J}$ is a homomorphism. Moreover, $\sharp = J \circ \bar{J}$, and since $\sharp$ is monomorphic, $\bar{J}$ also is monomorphic.

$\square$

Let $\Gamma \subset \operatorname{Aut}(\Lambda^S)$ be a super-Fuchsian group. The quotient set $\Lambda^S / \Gamma$ is called an $N = 1$ *super-Riemann surface* (or a *Riemann supersurface*). Two supersurfaces $P_1 = \Lambda^S / \Gamma_1$ and $P_2 = \Lambda^S / \Gamma_2$ are identified if $\Gamma_1$ and $\Gamma_2$ are conjugate in $\operatorname{Aut}(\Lambda^S)$. The projections $\sharp \colon \Lambda^S \to \Lambda$ and $\sharp \colon \Gamma \to \Gamma^\sharp$ determine the projection $\sharp \colon P \to P^\sharp = \Lambda / \Gamma^\sharp$.

By Theorem 7.2, the lifting $J^*$ determines an Arf function

$$\omega_P \colon H_1(P^\sharp, \mathbb{Z}_2) \to \mathbb{Z}_2.$$

We call the topological type $(g, \delta, k_\alpha, m_\alpha)$ of this Arf function the *topological type* of $P$.

THEOREM 11.1. *Let $Q$ be a Riemann surface and let $\omega \colon H_1(Q, \mathbb{Z}_2) \to \mathbb{Z}_2$ be an arbitrary Arf function. Then there is a super-Riemann surface $P$ such that $P^\sharp = Q$ and $\omega_P = \omega$.*

Proof. Let $Q = \Lambda/\Gamma$, where $\Gamma$ is a Fuchsian group. Then, by Theorem 7.2, there is a lifting $J^* \colon \Gamma \to \mathrm{SL}(2, \mathbb{R})$ generating the Arf function $\omega$. Now associate to an automorphism $\gamma \in \Gamma \subset \mathrm{Aut}(\Lambda)$ the automorphism $\gamma^S \in \Gamma \subset \mathrm{Aut}(\Lambda^S)$ of the form

$$\gamma^S(z|\theta) = \left( \frac{az+b}{cz+d} \,\middle|\, \frac{\theta}{cz+d} \right), \quad \text{where} \quad \begin{pmatrix} a & b \\ c & d \end{pmatrix} = J^*(\gamma).$$

Then the set

$$\Gamma^S = \{\gamma^S | \gamma \in \Gamma\} \subset \mathrm{Aut}(\Lambda^S)$$

is a group. Moreover, $(\gamma^S)^\sharp = \gamma$, whence $\Gamma^S$ is a super-Fuchsian group. We set $P = \Lambda^S/\Gamma^S$. Then $P^\sharp = Q$ and $\omega_P = \omega$. $\quad\square$

By Theorem 11.1, the set $T^t$ of super-Riemann surfaces of type

$$t = (g, \delta, k_\alpha, m_\alpha)$$

is nonempty for each $t$ satisfying the assumptions of Lemma 8.1. Our next goal is to describe the sets $T^t$.

## 12. Moduli of super-Riemann surfaces

**1.** The simplest superhyperbolic and superparabolic automorphisms are, respectively, of the form

$$I_\lambda^\sigma(z|\theta) = (\lambda z | \sigma \sqrt{\lambda} \theta)$$

and

$$J_\mu^\sigma(z|\theta) = (z + \mu | \sigma\theta),$$

where $\sigma = \pm 1$, $\lambda, \mu \in L_0(\mathbb{R})$, $\lambda^\sharp > 1$, $\mu^\sharp > 0$.

An immediate calculation shows that each element

$$A \in \mathrm{Aut}_2(\Lambda^S)$$

such that $A^\sharp(\infty) \neq \infty$ has the form

$$A = I(\sigma, \lambda, \alpha, \beta | \varepsilon, \delta),$$

where $\lambda, \alpha, \beta \in L_0(\mathbb{R})$, $\varepsilon, \delta \in L_1(\mathbb{R})$, $\sigma = \pm 1$, $\lambda^\sharp > 1$, $\alpha^\sharp \neq \beta^\sharp$ and

$$A(z|\theta) = \left( \frac{(\lambda\alpha - \beta)z + (1-\lambda)\alpha\beta}{(\lambda-1)z + (\alpha - \lambda\beta)} \right.$$
$$\left. - \frac{\lambda(\alpha-\beta)^2(\varepsilon + \delta z)\theta}{((\lambda-1)z + (\alpha - \lambda\beta))^2} \,\middle|\, \frac{\sigma\sqrt{\lambda}(\alpha - \beta)}{(\lambda-1)z + (\alpha - \lambda\beta)} \left( \theta + \varepsilon + \delta z + \frac{1}{2}\varepsilon\delta\theta \right) \right).$$

Set

$$I(\sigma, \lambda, \infty, 0 | 0, 0) = I_\lambda^\sigma.$$

Similarly, each element

$$A \in \mathrm{Aut}_1(\Lambda^S)$$

such that $A^\sharp(\infty) \neq \infty$ has the form

$$A = J(\sigma, \lambda, \alpha \,|\, \varepsilon, \delta),$$

where $\lambda, a \in L_0(\mathbb{R})$, $\varepsilon, \delta \in L_1(\mathbb{R})$, $\sigma = \pm 1$, $\lambda^\sharp > 0$ and

$$A(z|\theta) = \left( \frac{(1 - \alpha\lambda)z + a^2\lambda}{-\lambda z + (1 + a\lambda)} \right.$$
$$\left. - \frac{(\varepsilon + \delta z)\theta}{(-\lambda z + (1 + az))^2} \,\middle|\, \frac{\sigma(\theta + \varepsilon + \delta z + (1/2)\varepsilon\delta\theta)}{-\lambda z + (1 + a\lambda)} \right).$$

We use this parametrization of $\mathrm{Aut}_1(\Lambda^S) \cup \mathrm{Aut}_2(\Lambda^S)$ in our construction of the superanalog of the Frike–Klein–Teichmüller space.

As before, let $\gamma_{g,n}$ be the group generated by the elements

$$v_{g,n} = \{a_i, b_i \ (i = 1, \ldots, g), c_i \ (i = g + 1, \ldots, n)\}$$

satisfying the defining relations

$$\prod_{i=1}^{g} [a_i, b_i] \prod_{i=g+1}^{n} c_i = 1.$$

Set $c_i = [a_i, b_i]$ $(i = 1, \ldots, g)$.

Suppose $n = g + k + m$, $6g + 3k + 2m > 6$ and let $t = (g, \delta, k_\alpha, m_\alpha)$ be a topological type of an Arf function on the surface of type $(g, k, m)$. By Lemma 8.1, this means that

(1) $k = k_0 + k_1$, $m = m_0 + m_1$, $k_1 + m_1 \equiv 0 \pmod 2$;

(2) $\delta = 0$ for $k_1 + m_1 > 0$.

Denote by $\tilde{T}^t$ the set of monomorphisms $\psi \colon v_{g,n} \to \mathrm{Aut}(\Lambda^S)$ such that

(1) $(\psi(v_{g,n}))^\sharp \in \tilde{T}_{g,k,m}$;

(2) the super-Riemann surface $P = \Lambda^S / \psi(\gamma_{g,n})$ induces on $P^\sharp$ an Arf function $\omega = \omega_P$ of type $t$;

(3) the natural projection $(\psi(\gamma_{g,n}))^\sharp \to H_1(P^\sharp, \mathbb{Z}_2)$ takes $v_{g,n}$ to the standard basis $\{a_i, b_i \ (i = 1, \ldots, g), c_i \ (i = g + 1, \ldots, n)\}$ such that $\omega(a_i) = \omega(b_i) = 0$ for $i > 1$, $\omega(a_1) = \omega(b_1) = \delta$ and $\omega(c_i) = 0$ if and only if either $g + 1 \leqslant i \leqslant g + k_0$ or $g + k + 1 \leqslant i \leqslant g + k + m_0$.

The correspondence $\psi \mapsto \psi^\sharp$, where $\psi^\sharp(d) = (\psi(d))^\sharp$, generates a mapping $\sharp \colon \tilde{T}^t \to \tilde{T}_{g,k,m}$. By Theorem 11.1, we have

$$\sharp(\tilde{T}^t) = \tilde{T}_{g,k,m}.$$

The group $\mathrm{Aut}(\Lambda^S)$ acts on $\tilde{T}^t$ by conjugations. Let us set

$$T^t = \tilde{T}^t / \mathrm{Aut}(\Lambda^S).$$

For any triple of points

$$(z_1|\theta_1), (z_2|\theta_2), (z_3|\theta_3) \in \mathbb{R}^{(1|1)}$$

there is an element $h \in \mathrm{Aut}(\Lambda^S)$ such that

$$h(z_1|\theta_1) = (\infty|0), \quad h(z_2|\theta_2) = (0|0) \quad \text{and} \quad h(z_3|\theta_3) = (1|\theta).$$

Besides, automorphisms preserving the points $(\infty|0)$, $(0|0)$ and taking the points of the form $(1|\theta)$ to the points of the same form coincide with the involution $(z|\theta) \mapsto (z|-\theta)$. This fact allows one to identify $T^t$ with the set $\check{T}^t/\mathbb{Z}_2$, where

(1) for $g = k = 0$

$$\check{T}^t = \{\psi \in \tilde{T}^t \,|\, \psi(c_1 c_2) = I_\chi^\sigma, \psi(c_1) = J(\sigma_1, \lambda_1, -1 | \varepsilon, \delta)\};$$

(2) for $g = 0, k = 1$

$$\check{T}^t = \{\psi \in \tilde{T}^t \,|\, \psi(c_1) = I_{\lambda_1}^{\sigma_1}, \psi(c_2) = J(\sigma_2, \lambda_2, 1 | \varepsilon, \delta)\};$$

(3) for $g = k = 1$

$$\check{T}^t = \{\psi \in \tilde{T}^t \,|\, \psi(c_1) = I_{\lambda_1}^{\sigma_1}, \psi(b_1) = I(\sigma, \lambda_2, 1, \beta | \varepsilon, \delta)\};$$

(4) otherwise,

$$\check{T}^t = \{\psi \in \tilde{T}^t \,|\, \psi(c_1) = I_{\lambda_1}^{\sigma_1}, \psi(c_2) = I(\sigma_2, \lambda_2, 1, \beta | \varepsilon, \delta)\}.$$

In this notation, the numbers $\sigma, \sigma_1, \sigma_2$ are uniquely determined by the type $t$.

Now let us describe the mapping

$$\Psi : \check{T}^t \to \mathbb{R}^{(6g+3k+2m-6|4g+2k+2m-4)}.$$

To this end, we represent the elements $\psi(v_{g,m})$, where $\psi \in \check{T}^t$, in the following form:

$$\begin{aligned}
\psi(c_i) &= I(\sigma_i, \lambda_i, \alpha_i, \beta_i | x_i, y_i) && \text{for } i \leqslant g + k, \\
\psi(c_i) &= J(\sigma_i, \lambda_i, a_i | x_i, y_i) && \text{for } i > g + k, \\
\psi(a_i) &= I(\sigma_i^a, \lambda_i^a, \delta_i, \eta_i | x_i^a, y_i^a), \\
\psi(b_i) &= I(\sigma_i^b, \lambda_i^b, \varepsilon_i, \xi_i | x_i^b, y_i^b).
\end{aligned}$$

For $g = k = 0$ we set

$$\Psi(\psi) = \{\lambda_1, \lambda_3, \ldots, \lambda_{m-1}, a_3, \ldots, a_{m-2} | x_1, y_1, x_3, y_3, \ldots, x_{m-1}, y_{m-1}\}$$
$$\subset \mathbb{R}^{(2m-6|2m-4)}.$$

For $g = 0$, $k = 1$ we set

$$\Psi(\psi) = \{\lambda_1, \lambda_2, \ldots, \lambda_m, a_3, \ldots, a_{m-1} | x_2, y_2, \ldots, x_m, y_m\} \subset \mathbb{R}^{(2m-3|2m-2)}.$$

For $g = k = 1$ we set

$$\Psi(\psi) = \{\lambda_1, \ldots, \lambda_{m+1}, \eta_1, \varepsilon_1, \alpha_2, \beta_2, a_3, a_4, \ldots, a_m | $$
$$x_2, y_2, \ldots, x_m, y_m, x_1^a, y_1^a, x_1^b, y_1^b\}$$
$$\subset \mathbb{R}^{(2m+3|2m+2)}.$$

In the other cases, for $m = 0$ we set

$$\Psi(\psi) = \{\lambda_1, \ldots, \lambda_{g+k-1}, \xi_1, \eta_1, \varepsilon_1, \ldots, \xi_g, \eta_g, \varepsilon_g, \beta_2, \alpha_3, \beta_3, \ldots,$$
$$\alpha_{g+k-1}, \beta_{g+k-1} | x_2, y_2, \ldots, x_{g+k-1}, y_{g+k-1}, x_1^a, y_1^a, \ldots, x_g^a, y_g^a\}$$
$$\subset \mathbb{R}^{(6g+3k-6|4g+2k-4)},$$

for $m = 1$ we set

$$\Psi(\psi) = \{\lambda_1, \ldots, \lambda_{g+k}, \xi_1, \eta_1, \varepsilon_1, \ldots, \xi_g, \eta_g, \varepsilon_g, \beta_2, \alpha_3, \beta_3, \ldots,$$
$$\alpha_{g+k-1}, \beta_{g+k-1}, \alpha_{g+k} | x_2, y_2, \ldots, x_{g+k}, y_{g+k}, x_1^a, y_1^a, \ldots, x_g^a, y_g^a\}$$
$$\subset \mathbb{R}^{(6g+3k-4|4g+2k-2)},$$

and otherwise we set

$$\Psi(\psi) = \{\lambda_1, \ldots, \lambda_{g+k+m-1}, \xi_1, \eta_1, \varepsilon_1, \ldots, \xi_g, \eta_g, \varepsilon_g, \beta_2, \alpha_3, \beta_3, \ldots,$$
$$\alpha_{g+k}, \beta_{g+k}, a_{g+k+1}, \ldots, a_{g+k+m-2} | x_2, y_2, \ldots,$$
$$x_{g+k+m-1}, y_{g+k+m-1}, x_1^a, y_1^a, \ldots, x_g^a, y_g^a\}$$
$$\subset \mathbb{R}^{(6g+3k+2m-6|4g+2k+2m-4)}.$$

Set

$$\hat{T}^t = \Psi(\check{T}^t).$$

We say that a domain $Q \subset \mathbb{R}^{(r|\ell)}$ is *strongly diffeomorphic* to $\mathbb{R}^{(r|\ell)}$ if $Q^\sharp$ is diffeomorphic to $\mathbb{R}^r$ and $Q = \sharp^{-1}(\sharp(Q))$.

THEOREM 12.1 ([**62**], [**64**], [**73**]). *The mapping* $\Psi \colon \check{T}^t \to \hat{T}^t$ *is one-to-one and* $\hat{T}^t$ *is strongly diffeomorphic to*

$$\mathbb{R}^{(6g+3k+2m-6|4g+2k+2m-4)}.$$

Proof. The type $t$ determines the numbers $\sigma_i, \sigma_i^a, \sigma_i^b$ uniquely. The parameters $\lambda_i, \alpha_i, \beta_i, a_i, x_i, y_i$ uniquely determine $\psi(c_i)$ for $i < n$, whence the mapping

$$\psi(c_n) = \left(\prod_{i=1}^{n-1} \psi(c_i)\right)^{-1}.$$

This fact allows us to compute, by means of Eqs. (6)–(9), the parameters $\lambda_i^a, \lambda_i^b$ and $\delta_i$, which, together with $x_i^a, y_i^a, x_i, y_i$, uniquely determine $\psi(a_i)$ and $\psi(b_i)$. Hence the mapping $\Psi$ is one-to-one. All the restrictions on the parameters are the restrictions on their numerical parts. Thus,

$$\hat{T}^t = \sharp^{-1}(\sharp(\hat{T}^t)).$$

Moreover, by Theorem 4.1,

$$(\hat{T}^t)^\sharp = \hat{T}_{g,k,m} \cong \mathbb{R}^{6g+3m+2k-6}. \qquad \square$$

Our construction immediately implies

COROLLARY 12.1 ([**74**]). *Let $t$ be a topological type of an Arf function on a surface of type $(g, k, m)$ and let $M^t$ be the set of all super-Riemann surfaces of type $t$. Then $M^t$ can be represented in the form $T^t / \operatorname{Mod}^t$, where $T^t$ is strongly isomorphic to*

$$\mathbb{R}^{(6g+3k+2m-6|4g+2k+2m-4)}/\mathbb{Z}_2,$$

*and $\operatorname{Mod}^t$ acts on $T^t$ discretely. If $\omega$ is an Arf function of type $t$, then*

$$\operatorname{Mod}^t \cong \{\alpha \in \operatorname{Mod}_{g,k,m} \mid \omega\alpha = \omega\}.$$

**Example.** Consider the space $M_g$ of super Riemann surfaces $P$ such that $P^\sharp$ is the compact surface of genus $g > 1$. There are exactly two types of Arf functions on compact surfaces of genus $g > 0$. Hence $M_g$ decomposes into two connected components, each of which is diffeomorphic to

$$\mathbb{R}^{(6g-6|4g-4)}/\operatorname{Mod},$$

where Mod is a discrete group depending on the component.

## 13. $N = 2$ super-Fuchsian groups.
## $N = 2$ super-Riemann surfaces and their topological invariants

Denote by $A[a, b, c, d, \ell|\varepsilon]$ the mapping $A\colon \Lambda^{2S} \to \Lambda^{2S}$ of the form

$$A(z|\theta_1, \theta_2) = \left(\frac{az + b + \delta^{11}\theta_1 + \delta^{12}\theta_2}{cz + d + \delta^{21}\theta_1 + \delta^{22}\theta_2}\right|$$
$$\left.\frac{\ell^{11}\theta_1 + \ell^{12}\theta_2 + \varepsilon^{11}z + \varepsilon^{12}}{cz + d + \delta^{21}\theta_1 + \delta^{22}\theta_2}, \frac{\ell^{21}\theta_1 + \ell^{22}\theta_2 + \varepsilon^{21}z + \varepsilon^{22}}{cz + d + \delta^{21}\theta_1 + \delta^{22}\theta_2}\right),$$

where $a, b, c, d \in L_0(\mathbb{R})$, $\ell \in \operatorname{GL}(2, L_0(\mathbb{R}))$, $\varepsilon^{ij}, \delta^{ij} \in L_1(\mathbb{R})$.

According to [**45**], the automorphism group $\operatorname{Aut}(\Lambda^{2S})$ of the superdomain $\Lambda^{2S}$ consists of transformations $A[a, b, c, d, \ell|\varepsilon]$ such that

$$\begin{pmatrix} -c & a \\ -d & b \end{pmatrix}\begin{pmatrix} \delta^{11} & \delta^{12} \\ \delta^{21} & \delta^{22} \end{pmatrix} = \begin{pmatrix} \varepsilon^{21} & \varepsilon^{11} \\ \varepsilon^{22} & \varepsilon^{12} \end{pmatrix}\begin{pmatrix} \ell^{11} & \ell^{12} \\ \ell^{21} & \ell^{22} \end{pmatrix}$$

and

$$ad - bc - \varepsilon^{11}\varepsilon^{12} - \varepsilon^{21}\varepsilon^{22} = \ell^{11}\ell^{22} + \ell^{21}\ell^{12} + \delta^{11}\delta^{22} + \delta^{12}\delta^{21} = \Delta,$$

where $\Delta^\sharp > 0$,

$$\ell^{11}\ell^{21} + \delta^{11}\delta^{21} = \ell^{12}\ell^{22} + \delta^{12}\delta^{22} = 0.$$

An immediate calculation easily leads to

LEMMA 13.1. *Any automorphism $A[a, b, c, d, \ell|\varepsilon]$ belongs to one of the following two types:*
(1) *(nontwisted) $(\ell^{12})^\sharp = (\ell^{21})^\sharp = 0$;*
(2) *(twisted) $(\ell^{11})^\sharp = (\ell^{22})^\sharp = 0$.*
*A nontwisted (respectively, twisted) automorphism is uniquely determined by the parameters $a$, $b$, $c$, $d$, $\varepsilon^{ij}$, $\ell^{11}$ (respectively, $a$, $b$, $c$, $d$, $\varepsilon^{ij}$, $\ell^{12}$). These parameters can take arbitrary values under the restrictions $a, b, c, d, \ell^{ij} \in L_0(\mathbb{R})$, $\varepsilon^{ij} \in L_1(\mathbb{R})$, $(ad - bc)^\sharp > 0$, $(\ell^{11} + \ell^{12})^\sharp \neq 0$.*

LEMMA 13.2. *Let $(z_i | \theta_i^1, \theta_i^2) \in \mathbb{R}^{(1|2)}$, $i = 1, 2, 3$, and suppose $z_i^\sharp \neq z_j^\sharp$ for $i \neq j$. Then there is an automorphism $A \in \mathrm{Aut}(\Lambda^{2S})$ such that $A(z_1 | \theta_1^1, \theta_1^2) = (0|0,0)$, $A(z_2 | \theta_2^1, \theta_2^2) = (\infty | 0, 0)$ and $A(z_3 | \theta_3^1, \theta_3^2) = (1 | \theta^1, \theta^2)$.*

Set

$$\mathrm{Aut}_0^{2t} = \{ h \in \mathrm{Aut}(\Lambda^{2S}) \mid h(0|0,0) = (0|0,0),$$

$$h(\infty|0,0) = (\infty|0,0), h(1|0,0) = (1|\theta^1, \theta^2) \}.$$

LEMMA 13.3. *The group $A_0^{2S}$ consists of elements of the form $h(z|\theta^1, \theta^2) = (z | \mu\theta^i, \mu^{-1}\theta^{3-i})$, where $\mu \in L_0(\mathbb{R})$, $\mu^\sharp \neq 0$ and $i$ is either $1$ or $2$.*

Proof. The definition of $A_0^{2S}$ implies $h = A[1, 0, 0, 1, \ell | 0]$, whence $\ell^{11}\ell^{21} = \ell^{12}\ell^{22} = 0$, $\ell^{11}\ell^{22} + \ell^{12}\ell^{21} = 1$. Therefore, either $\ell^{12} = \ell^{21} = 0$, $\ell^{11}\ell^{22} = 1$ or $\ell^{11} = \ell^{22} = 0$, $\ell^{12}\ell^{21} = 1$. $\qquad\square$

The correspondence $A \mapsto A^\sharp$, where

$$A = A[a, b, c, d, \ell | \varepsilon], \qquad A^\sharp(z) = \frac{a^\sharp z + b^\sharp}{c^\sharp z + d^\sharp},$$

induces an epimorphism $\sharp\colon \mathrm{Aut}(\Lambda^{2S}) \to \mathrm{Aut}(\Lambda)$. An element of the set $\mathrm{Aut}_1(\Lambda^{2S}) = (\sharp)^{-1}(\mathrm{Aut}_1(\Lambda))$ is said to be *superparabolic*, and an element of the set $\mathrm{Aut}_2(\Lambda^{2S}) = (\sharp)^{-1}(\mathrm{Aut}_2(\Lambda))$ is said to be *superhyperbolic*.

A subgroup $\Gamma \subset \mathrm{Aut}(\Lambda^{2S})$ is said to be $N = 2$ *super-Fuchsian* if $\Gamma^\sharp = \sharp(\Gamma)$ is a Fuchsian group and $\sharp\colon \Gamma \to \Gamma^\sharp$ is an isomorphism. In this section we study only $N = 2$ super-Fuchsian groups consisting of superparabolic and superhyperbolic automorphisms of $\Lambda^{2S}$.

The relations on the numbers $\ell^{ij}$ imply that $(\ell^{11}\ell^{22})^\sharp \geqslant 0$, $(\ell^{12}\ell^{21})^\sharp \geqslant 0$ and either $(\ell^{11})^\sharp = (\ell^{22})^\sharp = 0$ or $(\ell^{12})^\sharp = (\ell^{21})^\sharp = 0$.

Associate to an automorphism $A = A[a, b, c, d, \ell | \varepsilon]$ the matrix

$$\bar{J}(A) = \frac{\sigma}{\sqrt{a^\sharp d^\sharp - b^\sharp c^\sharp}} \begin{pmatrix} a^\sharp & b^\sharp \\ c^\sharp & d^\sharp \end{pmatrix} \in \mathrm{SL}(2, \mathbb{R}),$$

where $\sigma = \mathrm{sign}(\ell^{11} + \ell^{12} + \ell^{21} + \ell^{22})^\sharp$.

LEMMA 13.4. *Let $\Gamma \in \mathrm{Aut}(\Lambda^{2S})$ be an $N = 2$ super-Fuchsian group. Then the correspondence $\bar{J}\colon \Gamma \to \mathrm{SL}(2, \mathbb{R})$ is a monomorphism and, therefore, determines a lifting $J^*\colon \Gamma^\sharp \to \bar{\Gamma} = \bar{J}(\Gamma)$.*

Proof. A direct calculation shows that $\bar{J}$ is a homomorphism. Besides, $\sharp = J \circ \bar{J}$, and since $\sharp$ is monomorphic, $\bar{J}$ also is monomorphic. $\qquad\square$

Let $\Gamma \subset \mathrm{Aut}(\Lambda^{2S})$ be an $N = 2$ super-Fuchsian group. The quotient $\Lambda^{2S}/\Gamma$ is called an $N = 2$ *super-Riemann surface* or an $N = 2$ *Riemann supersurface*. Two $N = 2$ super-Riemann surfaces $P_1 = \Lambda^{2S}/\Gamma_1$ and $P_2 = \Lambda^{2S}/\Gamma_2$ are considered coinciding if $\Gamma_1$ and $\Gamma_2$ are conjugate in $\mathrm{Aut}(\Lambda^{2S})$. The projections $\sharp\colon \Lambda^{2S} \to \Lambda$ and $\sharp\colon \Gamma \to \Gamma^\sharp$ determine a projection $\sharp\colon P \to P^\sharp = \Lambda/\Gamma^\sharp$.

By Theorem 7.2, the lifting $J^*$ determines an Arf function

$$\omega_P^1 \colon H_1(P^\sharp, \mathbb{Z}_2) \to \mathbb{Z}_2.$$

Now consider the function

$$\Omega \colon \Gamma \to \mathbb{Z}_2 = \{0, 1\}$$

that takes an automorphism $A[a, b, c, d, \ell | \varepsilon] \in \Gamma$ to 0 if $(\ell^{12})^\sharp = (\ell^{21})^\sharp = 0$, and to 1 if $(\ell^{11})^\sharp = (\ell^{22})^\sharp = 0$. It is easy to see that $\Omega$ is a homomorphism and that it induces a homomorphism $\omega_P^0 \colon H_1(P^\sharp, \mathbb{Z}_2) \to \mathbb{Z}_2$. Denote by $\omega_P^2$ the Arf function $\omega_P^1 + \omega_P^0$.

THEOREM 13.1. *Let $Q$ be a Riemann surface and let $\omega^1, \omega^2 \colon H_1(Q, \mathbb{Z}_2) \to \mathbb{Z}_2$ be arbitrary Arf functions. Then there is an $N = 2$ super-Riemann surface $P$ such that $P^\sharp = Q$ and $\omega_P^1 = \omega^1$, $\omega_P^2 = \omega^2$.*

Proof. Let $Q = \Lambda/\Gamma$, where $\Gamma$ is a Fuchsian group. Then, by Theorem 7.2, there is a lifting $J^* \colon \Gamma \to \mathrm{SL}(2, \mathbb{R})$ generating the Arf function $\omega^1$. Now let us associate to an automorphism $\gamma \in \Gamma \subset \mathrm{Aut}(\Lambda)$ the automorphism $\gamma^S \in \mathrm{Aut}(\Lambda^{2S})$ of the form

$$\gamma^S(z|\theta_1, \theta_2) = \left( \frac{az + b}{cz + d} \,\middle|\, \frac{\theta_i}{cz + d}, \frac{\theta_{3-i}}{cz + d} \right),$$

where $\left( \begin{smallmatrix} a & b \\ c & d \end{smallmatrix} \right) = J^*(\gamma)$ and

$$i = \begin{cases} 1 & \text{if } \omega^1(\gamma) = \omega^2(\gamma), \\ 2 & \text{if } \omega^1(\gamma) \neq \omega^2(\gamma). \end{cases}$$

Since $\omega^1 + \omega^2$ is a homomorphism, the group $\Gamma^S = \{\gamma^S | \gamma \in \Gamma\}$ is an $N = 2$ super-Fuchsian group and $P = \Lambda^{2S}/\Gamma^S$ is an $N = 2$ super-Riemann surface. Here $P^\sharp = Q$, $\omega_P^1 = \omega^1$ and $\omega_P^2 = \omega^2$. □

We say that an $N = 2$ super-Riemann surface $P$ is *nontwisted* if $\omega_P^0 = 0$. Its *topological type* is the topological type $(g, \delta, k_\alpha, m_\alpha)$ of the Arf function $\omega_P^1 = \omega_P^2$. If $\omega_P^0 \neq 0$, then we say that the $N = 2$ super-Riemann surface is *twisted*. Its *topological type* is the topological type $(g, \delta_1, \delta_2, k_{\alpha\beta}, m_{\alpha\beta})$ of the pair of Arf function $(\omega_P^1, \omega_P^2)$.

Our next problem is to describe the sets $M^{2t}$ of $N = 2$ super-Riemann surfaces of arbitrary topological type $t$.

## 14. Moduli of $N = 2$ super-Riemann surfaces

The simplest superhyperbolic and superparabolic automorphisms of the superdomain $\Lambda^{2S}$ have the form

$$I_\lambda^\ell(z|\theta_1, \theta_2) = (\lambda z | \ell^{11}\theta_1 + \ell^{12}\theta_2, \ell^{21}\theta_1 + \ell^{22}\theta_2)$$

and

$$J_\mu^\ell(z|\theta_1, \theta_2) = (z + \mu | \ell^{11}\theta_1 + \ell^{12}\theta_2, \ell^{21}\theta_1 + \ell^{22}\theta_2),$$

respectively, where $\lambda, \mu \in L_0(\mathbb{R})$, $\lambda^\sharp > 1$, $\mu^\sharp > 0$, $\ell \in \mathrm{GL}(2, L_0(\mathbb{R}))$ and $\ell^{11}\ell^{21} = \ell^{12}\ell^{22} = 0$, $\ell^{11}\ell^{22} + \ell^{12}\ell^{21}$ is $\lambda$ for $I_\lambda^\ell$ and 1 for $I_\mu^\ell$.

Each automorphism $A \in \mathrm{Aut}_2(\Lambda^{2S})$ such that $A^\sharp(\infty) \neq \infty$ has the form

$$A = I(\ell, \lambda, \alpha, \beta \,|\, \varepsilon) = A[(\lambda\alpha - \beta), (1 - \alpha)\alpha\beta, (\lambda - 1), (\alpha - \lambda\beta), \ell \,|\, \varepsilon],$$

where $\lambda, \alpha, \beta \in L_0(\mathbb{R})$, $\lambda^\sharp > 1$, $\alpha^\sharp \neq \beta^\sharp$, $\ell \in \mathrm{GL}(2, L_0(\mathbb{R}))$, $\varepsilon \in \mathrm{GL}(2, L_1(\mathbb{R}))$. Let us set

$$I(\ell, \lambda, \infty, 0 \,|\, 0) = I_\lambda^\ell.$$

Each automorphism $A \in \mathrm{Aut}_1(\Lambda^{2S})$ such that $A^\sharp(\infty) \neq \infty$ has the form

$$A = J(\ell, \lambda, a \,|\, \varepsilon) = A[(1 - a\lambda), a^2\lambda, -\lambda, (1 + a\lambda), \ell \,|\, \varepsilon],$$

where $\lambda, a \in L_0(\mathbb{R})$, $\lambda^\sharp > 0$, $\ell \in \mathrm{GL}(2, L_0(\mathbb{R}))$, $\varepsilon \in \mathrm{GL}(2, L_1(\mathbb{R}))$.

Similarly to the case of $N = 1$ super-Riemann surfaces, we use this parametrization of $\mathrm{Aut}_1(\Lambda^{2S}) \cup \mathrm{Aut}_2(\Lambda^{2S})$ in the construction of the $N = 2$ superanalog of the Frike–Klein–Teichmüller space.

Denote by $\tilde{T}^{2t}$ the set of monomorphisms $\psi \colon v_{g,n} \to \mathrm{Aut}(\Lambda^{2S})$ such that
(1) $(\psi(v_{g,n}))^\sharp \in \tilde{T}_{g,k,m}$;
(2) $P = \Lambda^{2S}/\psi(\gamma_{g,n})$ is an $N = 2$ super-Riemann surface of type $t$;
(3) the natural projection $(\psi(\gamma_{g,n}))^\sharp \to H_1(P^\sharp, \mathbb{Z}_2)$ takes $v_{g,n}$ to a standard basis $\{a_i, b_i \ (i = 1, \dots, g), c_i \ (i = g + 1, \dots, n)\}$ such that the Arf functions $\omega_i = \omega_P^i$ satisfy the assumptions of Lemma 8.1 (if the supersurface is nontwisted) or of Lemma 8.2 (if it is twisted).

The correspondence $\psi \mapsto \psi^\sharp$, where $\psi^\sharp(d) = (\psi(d))^\sharp$, induces a mapping $\sharp \colon \tilde{T}^{2t} \to \tilde{T}_{g,k,m}$. By Theorem 13.1, we have $\sharp(\tilde{T}^{2t}) = \tilde{T}_{g,k,m}$.

The correspondence $\Psi \colon \psi \mapsto \Lambda^{2S}/\psi(\gamma_{g,n})$ induces a topology from $\tilde{T}^{2t}$ on the space $M_{2t}$ of $N = 2$ super-Riemann surfaces of topological type $t$.

Now consider the space $\check{T}^{2t}$ defined as follows:
(1) for $g = k = 0$

$$\check{T}^{2t} = \{\psi \in \tilde{T}^{2t} \,|\, \psi(c_1 c_2) = I_\lambda^\ell, \psi(c_1) = J(\ell_1, \lambda_1, -1 \,|\, \varepsilon)\};$$

(2) for $g = 0$, $k = 1$

$$\check{T}^{2t} = \{\psi \in \tilde{T}^{2t} \,|\, \psi(c_1) = I_{\lambda_1}^{\ell_1}, \psi(c_2) = J(\ell_2, \lambda_2, 1 \,|\, \varepsilon)\};$$

(3) for $g = k = 1$

$$\check{T}^{2t} = \{\psi \in \tilde{T}^{2t} \,|\, \psi(c_1) = I_{\lambda_1}^{\ell_1}, \psi(b_1) = I(\ell, \lambda_2, 1 \,|\, \varepsilon)\};$$

(4) otherwise, we have

$$\check{T}^{2t} = \{\psi \in \tilde{T}^{2t} \,|\, \psi(c_1) = I_{\lambda_1}^{\ell_1}, \psi(c_2) = I(\ell_2, \lambda_2, 1, \beta \,|\, \varepsilon)\}.$$

Now let us describe the mapping

$$\Psi^{2t} \colon \check{T}^{2t} \to \mathbb{R}^{(r_1 | r_2)}.$$

In order to do this, represent the elements $\psi(v_{g,m})$, where $\psi \in T^{2t}$, in the following form:

$$\psi(c_i) = I(\ell_i, \lambda_i, \alpha_i, \beta_i \,|\, z_i) \qquad \text{for } i \leqslant g + k,$$
$$\psi(c_i) = J(\ell_i, \lambda_i, a_i \,|\, z_i) \qquad \text{for } i > g + k,$$
$$\psi(a_i) = I(\ell_i^a, \lambda_i^a, \delta_i, \eta_i \,|\, z_i^a),$$
$$\psi(b_i) = I(\ell_i^b, \lambda_i^b, \varepsilon_i, \xi_i \,|\, z_i^b).$$

By Lemma 13.1, for a given topological type $t$, the numbers $\ell, \ell_i, \ell_i^a, \ell_i^b$ are determined by the parameters

$$\tilde{\ell}, \tilde{\ell}_i, \tilde{\ell}_i^a, \tilde{\ell}_i^b \in \Lambda_0(\mathbb{R}),$$

where

$$\tilde{\ell}, \tilde{\ell}_i, \tilde{\ell}_i^a, \tilde{\ell}_i^b = (\pm\ell, \pm\ell_i, \pm\ell_i^a, \pm\ell_i^b),$$

with positive numerical parts $\tilde{\ell}^\sharp, \tilde{\ell}_i^\sharp, (\tilde{\ell}_i^a)^\sharp, (\tilde{\ell}_i^b)^\sharp > 0$.

For a nontwisted topological type $t$ and $k = m = 0$ let

$$\Psi(\psi) = \{\lambda_1, \ldots, \lambda_{g-1}, \tilde{\ell}_1^a, \tilde{\ell}_1^b, \ldots, \tilde{\ell}_g^a, \tilde{\ell}_g^b, \xi_1, \eta_1, \varepsilon_1, \ldots, \xi_g, \eta_g, \varepsilon_g,$$
$$\beta_2, \alpha_3, \beta_3, \ldots, \alpha_{g-1}, \beta_{g-1} \,|\, z_2, \ldots, z_{g-1}, z_1^a, \ldots, z_g^a\}$$
$$\subset \mathbb{R}^{(8g-6 \,|\, 8g-8)}.$$

For a twisted topological type $t$ and $k = m = 0$ let

$$\Psi(\psi) = \{\lambda_1, \ldots, \lambda_{g-1}, \tilde{\ell}_1^b, \tilde{\ell}_2^a, \tilde{\ell}_2^b, \ldots, \tilde{\ell}_g^a, \tilde{\ell}_g^b, \xi_1, \eta_1, \varepsilon_1, \ldots, \xi_g, \eta_g, \varepsilon_g,$$
$$\beta_2, \alpha_3, \beta_3, \ldots, \alpha_{g-1}, \beta_{g-1} \,|\, z_2, \ldots, z_{g-1}, z_1^a, \ldots, z_g^a\}$$
$$\subset \mathbb{R}^{(8g-7 \,|\, 8g-8)}.$$

For $g = k = 0$ let

$$\Psi(\psi) = \{\tilde{\ell}, \tilde{\ell}_1, \tilde{\ell}_3, \ldots, \tilde{\ell}_{m-1}, \lambda_1, \lambda_3, \ldots, \lambda_{m-1}, a_3, \ldots, a_{m-2} \,|\, z_1, z_3, \ldots, z_{m-1}\}$$
$$\subset \mathbb{R}^{(3m-7 \,|\, 4m-8)}.$$

For $g = 0$, $k = 1$ let

$$\Psi(\psi) = \{\tilde{\ell}_1, \tilde{\ell}_2, \ldots, \tilde{\ell}_m, \lambda_1, \lambda_2, \ldots, \lambda_m, a_3, \ldots, a_{m-1} \,|\, z_2, \ldots, z_m\}$$
$$\subset \mathbb{R}^{(3m-3 \,|\, 4m-4)}.$$

For $g = k = 1$ let

$$\Psi(\psi) = \{\tilde{\ell}_2, \ldots, \tilde{\ell}_{m+1}, \lambda_1, \ldots, \lambda_{m+1}, \tilde{\ell}_1^a, \tilde{\ell}_1^b, \eta_1, \varepsilon_1,$$
$$\alpha_2, \beta_2, a_3, \ldots, a_m \,|\, z_2, \ldots, z_m, z_1^a, z_1^b\}$$
$$\subset \mathbb{R}^{(3m+5 \,|\, 4m+4)}.$$

For $m = 0$ let

$$\Psi(\psi) = \{\tilde{\ell}_{g+1}, \ldots, \tilde{\ell}_{g+k-1}, \lambda_1, \ldots, \lambda_{g+k-1}, \tilde{\ell}_1^a, \tilde{\ell}_1^b, \ldots, \tilde{\ell}_g^a \tilde{\ell}_g^b, \xi_1, \eta_1, \varepsilon_1, \ldots,$$
$$\xi_g, \eta_g, \varepsilon_g, \beta_2, \alpha_3, \beta_3, \ldots, \alpha_{g+k-1}, \beta_{g+k-1} | z_2, \ldots, z_{g+k-1}, z_1^a, \ldots, z_g^a\}$$
$$\subset \mathbb{R}^{(8g+4k-7|8g+4k-8)}.$$

For $m = 1$ let

$$\Psi(\psi) = \{\tilde{\ell}_{g+1}, \ldots, \tilde{\ell}_{g+k}, \lambda_1, \ldots, \lambda_{g+k}, \tilde{\ell}_1^a, \tilde{\ell}_1^b, \ldots, \tilde{\ell}_g^a \tilde{\ell}_g^b, \xi_1, \eta_1, \varepsilon_1, \ldots, \xi_g,$$
$$\eta_g, \varepsilon_g, \beta_2, \alpha_3, \beta_3, \ldots, \alpha_{g+k-1}, \beta_{g+k-1}, \alpha_{g+k} | z_2, \ldots, z_{g+k}, z_1^a, \ldots, z_g^a\}$$
$$\subset \mathbb{R}^{(8g+4k-4|8g+4k-4)}.$$

For $m > 1$ let

$$\Psi(\psi) = \{\tilde{\ell}_{g+1}, \ldots, \tilde{\ell}_{g+k+m-1}, \lambda_1, \ldots, \lambda_{g+k+m-1}, \tilde{\ell}_1^a, \tilde{\ell}_1^b, \ldots, \tilde{\ell}_g^a \tilde{\ell}_g^b,$$
$$\xi_1, \eta_1, \varepsilon_1, \ldots, \xi_g, \eta_g, \varepsilon_g, \beta_2, \alpha_3, \beta_3, \ldots, \alpha_{g+k}, \beta_{g+k},$$
$$a_{g+k+1}, \ldots, a_{g+k+m-2} | z_2, \ldots, z_{g+k+m-1}, z_1^a, \ldots, z_g^a\}$$
$$\subset \mathbb{R}^{(8g+4k+3m-7|8g+4k+4m-8)}.$$

Now we set

$$\hat{T}^{2t} = \Psi(\check{T}^{2t}).$$

THEOREM 14.1. *The mapping* $\Psi \colon \check{T}^{2t} \to \hat{T}^{2t}$ *is one-to-one and the space* $\hat{T}^{2t}$ *is strongly diffeomorphic to*

$$\mathbb{R}^{(8g+4k+3m-b|8g+4k+4m-8)},$$

*where* (1) $b = 6$ *if* $k = m = 0$ *and* $t$ *is a nontwisted topological type*; (2) $b = 7$ *otherwise.*

The proof is similar to that of Theorem 12.1.

The group $\text{Aut}(\Lambda^{2S})$ acts on $\tilde{T}^{2t}$ by conjugations. Let us set

$$T^{2t} = \tilde{T}^{2t} / \text{Aut}(\Lambda^{2S}).$$

For an arbitrary triple

$$(z_1 | \theta_1^1, \theta_1^2), (z_2 | \theta_2^1, \theta_2^2), (z_3 | \theta_3^1, \theta_3^2) \in \mathbb{R}^{(1|2)}$$

there is an automorphism $h \in \text{Aut}(\Lambda^{2S})$ such that

$$h(z_1 | \theta_1^1, \theta_1^2) = (\infty | 0, 0), \quad h(z_2 | \theta_2^1, \theta_2^2) = (0 | 0, 0), \quad h(z_3 | \theta_3^1, \theta_3^2) = (1 | \theta^1, \theta^2).$$

Hence, to each point in $T^{2t}$ at least one point in $\check{T}^{2t}$ is associated. Besides, the group

$$\widehat{\text{Aut}}(\Lambda^{2S}) = \{h \in \text{Aut}(\Lambda^{2S}) | h(\check{T}^{2t}) = \check{T}^{2t}\}$$

consists of mappings taking points of the form $(\infty | 0, 0)$, $(0 | 0, 0)$, $(1 | \theta^1, \theta^2)$ to points of the same form, and therefore, it is generated by the mappings

$$h_\lambda \colon (z | \theta^1, \theta^2) \mapsto (z | \lambda \theta^1, \lambda^{-1} \theta^2),$$
$$h^1 \colon (z | \theta^1, \theta^2) \mapsto (z | \theta^2, \theta^1),$$

and
$$h^2 \colon (z \,|\, \theta^1, \theta^2) \mapsto (z \,|\, -\theta^1, -\theta^2).$$

The group of mappings of the form $h_\lambda$ acts on $\tilde{T}^{2t}$ faithfully if $t$ is a type of twisted pairs of Arf functions, and it acts on $\tilde{T}^{2t}$ identically if $t$ is a type of nontwisted Arf-functions.

By Theorem 14.1, this implies the following statement.

THEOREM 14.2 ([**72**], [**74**]). *The set $T^{2t}$ is strongly homeomorphic to*
$$\mathbb{R}^{(8g+4k+3m-b(t)\,|\,8g+4k+4m-8)}/(\mathbb{Z}_2)^2,$$
*where*: (1) $b(t) = 6$ *if $k = m = 0$ and $t$ is a type of a nontwisted pair of Arf functions*; (2) $b(t) = 8$ *if $t$ is a type of a twisted pair of Arf functions*; (3) $b(t) = 7$ *otherwise.*

Hence we obtain

COROLLARY 14.1. *Let $t$ be a topological type of Arf functions on a surface of type $(g, k, m)$ and let $M^{2t}$ be the set of all $N = 2$ super-Riemann surfaces of type $t$. Then $M^t$ can be represented in the form $T^{2t}/\operatorname{Mod}^{2t}$, where $T^{2t}$ is strongly diffeomorphic to*
$$\mathbb{R}^{(8g+4k+3m-b(+)\,|\,8g+4k+4m-8)}/(\mathbb{Z}_2)^2,$$
*and the group $\operatorname{Mod}^{2t}$ acts on $T^{2t}$ discretely. If $(\omega_1, \omega_2)$ is a pair of Arf functions of type $t$, then*
$$\operatorname{Mod}^{2t} \cong \{\alpha \in \operatorname{Mod}_{g,k,m} \mid \omega_1 \alpha = \omega_1, \omega_2 \alpha = \omega_2\}.$$

## 15. Superholomorphic morphisms of super-Riemann surfaces

**1.** Let $\tilde{U} = \{(\tilde{z}, \tilde{\theta})\}$ and $U = \{(z, \theta)\}$ be two superdomains in $\mathbb{C}^{(1|1)}$. Consider a mapping $f \colon U \to \tilde{U}$. It is said to be *superconformal* if
$$\tilde{z} = u(z) - u'(z)\varepsilon(z)\theta,$$
$$\tilde{\theta} = \sqrt{u'(z)}(\theta + \varepsilon(z) + \tfrac{1}{2}\varepsilon(z)\varepsilon'(z)\theta),$$
where $u(z) \in L_0(\mathbb{C})$, $\varepsilon(z) \in L_1(\mathbb{C})$ (see [**88**]).

According to [**45**], the group of superconformal automorphisms of $\Lambda^s$ coincides with $\operatorname{Aut}(\Lambda^s)$.

Now suppose $\Gamma \subset \operatorname{Aut}(\Lambda^s)$ is a super-Fuchsian group, and $P = \Lambda^s/\Gamma$ is a super-Riemann surface; let $\phi \colon \Lambda^s \to P$ be the natural projection and let $\phi^\sharp \colon \Lambda \to P^\sharp$ be its numerical part. Consider a domain $U \subset P$, a connected component $V$ of the preimage $(\phi^\sharp)^{-1}(U^\sharp)$, and set $\tilde{U} = (\sharp)^{-1}(V)$. We call the set $U$ a *neighborhood* of a point $p \in U$ if $U = \phi(\tilde{U})$. If this is the case, then we call the mapping $\varepsilon \colon U \to \mathbb{C}^{(1|1)}$ a *local chart* if $\varepsilon\phi \colon \tilde{U} \to \mathbb{C}^{(1|1)}$ is superconformal and $\varepsilon^\sharp(p_1^\sharp) \neq \varepsilon^\sharp(p_2^\sharp)$ for $p_1^\sharp \neq p_2^\sharp$. A set of local charts $\{\varepsilon_i\}$ covering $P$ is called an *atlas* on $P$.

Now let $\tilde{P}$ be another super-Riemann surface and let $\{\tilde{\varepsilon}_i\}$ be an atlas on it. We call a mapping $f \colon \tilde{P} \to P$ a *superholomorphic morphism* if $f(\tilde{P}) = P$

and $f$ is superconformal in the atlases $\{\varepsilon_i\}$ and $\{\tilde{\varepsilon}_i\}$. An invertible super-holomorphic isomorphism is called a *superholomorphic isomorphism.*

Two superholomorphic morphisms $f_1\colon \tilde{P}_1 \to P_1$ and $f_2\colon \tilde{P}_2 \to P_2$ are considered the same if there are superholomorphic morphisms $\tilde{\omega}\colon \tilde{P}_1 \to \tilde{P}_2$ and $\omega\colon P_1 \to P_2$ such that $f_2\tilde{\phi} = \phi f_1$. We say that $f_1$ and $f_2$ have the same topological type if there are homeomorphisms $\tilde{\psi}^\sharp\colon \tilde{P}_1^\sharp \to \tilde{P}_2^\sharp$ and $\psi^\sharp\colon P_1^\sharp \to P_2^\sharp$ such that $f_2^\sharp \tilde{\psi}^\sharp = \psi^\sharp f_1^\sharp$ and $\omega_1 = \omega_2 \psi^\sharp$, where $\omega_i$ is the Arf function corresponding to $P_i$.

We call a superholomorphic morphism a *supercovering* [**69**] if $f^\sharp$ has no critical points. To each superholomorphic morphism $f$ a supercovering $f_0\colon P \setminus B \to P' \setminus B'$, where $B' = (\sharp)^{-1}(\{\text{the set of critical values of } f^\sharp\})$ is associated and $B = f^{-1}(B')$. Similarly to the classical case (see Section 6), a superholomorphic morphism is totally determined by the corresponding supercovering together with the indication which punctures of the covering $f^\sharp$ must be patched up. That is why in the sequel we consider only supercoverings.

**Remark.** In contrast to the classical case, not each puncture $p \in P^\sharp$ of a supercovering $f\colon \tilde{P} \to P$ can be patched up in the supercase. For this to be possible, it is necessary and sufficient that the value of the Arf function $\omega\colon H_1(P^\sharp, Z_2) \to Z_2$ at the puncture $p$ and the value of the Arf function $\tilde{\omega}\colon H_1(\tilde{P}^\sharp, Z_2) \to Z_2$ at the puncture $(f^\sharp)^{-1}(p)$ were both equal to 0. If the first of these assumptions is satisfied, then the second assumption is equivalent to the requirement that the ramification orders at the points $(f^\sharp)^{-1}(p)$ were odd.

**2.** Now let us describe classes of topological equivalence of supercoverings. Let $t$ be a topological type of an Arf function on a surface of type $(g, k, m)$. Consider a subgroup $\gamma \subset \gamma_{g,k+m} = \gamma_{g,n}$, and let $\psi \in \tilde{T}^t$ and $P = \Lambda^s/\psi(\gamma_{g,n})$. The Arf function corresponding to $P$ generates a function $\omega\colon \gamma_{g,n} \to Z_2$. We set

$$\widetilde{\mathrm{Mod}}^{(\gamma,\omega)} = \{\alpha \in \mathrm{Mod}_{g,k,m} \mid \omega\alpha = \omega, \alpha\gamma = \gamma\}$$

and

$$\mathrm{Mod}^{(\gamma,\omega)} = \widetilde{\mathrm{Mod}}^{(\gamma,\omega)}/(\widetilde{\mathrm{Mod}}^{(\gamma,\omega)} \cap \mathrm{IMod}_{g,k,m}).$$

Repeating the argument from Section 6 we obtain

THEOREM 15.1. *The set of supercoverings of a given topological type is a connected supermanifold of the form $T^t/\mathrm{Mod}^{(\gamma,\omega)}$. To each pair $(\gamma,\omega)$ a connected component of the space of superholomorphic coverings is associated. The two components corresponding to pairs $(\gamma,\omega)$ and $(\gamma',\omega')$ coincide if and only if they are taken to each other by the action of the group $\mathrm{Mod}_{g,k,m}$.*

# CHAPTER 2

# Moduli of Real Algebraic Curves and Their Superanalogs. Differentials, Spinors, and Jacobians of Real Curves

## 1. Topological types of real algebraic curves

**1.** A *real algebraic (nonsingular) curve* is a pair $X = (P, \tau)$ where $P = X(\mathbb{C})$ is a compact Riemann surface (called the *complexification*) and $\tau = \tau_X \colon P \to P$ is an antiholomorphic involution (called the *complex conjugation involution*). The fixed points $X(\mathbb{R}) = P^\tau$ of this involution form the set of real points of the curve. For example, a real plane algebraic curve $F(x, y) = 0$ corresponds to the pair $(P, \tau)$, where $P$ is the normalization of the compactified surface $\{(x, y) \in \mathbb{C}^2 \mid F(x, y) = 0\}$, and $\tau$ is induced by the involution $(x, y) \mapsto (\bar{x}, \bar{y})$.

Two real algebraic curves $X_1 = (P_1, \tau_1)$ and $X_2 = (P_2, \tau_2)$ are considered to be the same if there is a biholomorphic mapping $\psi \colon P_1 \to P_2$ such that $\psi \tau_1 = \tau_2 \psi$.

A curve $X$ is said to be *separating* (type I in Klein's classification) if the set $X(\mathbb{C}) \setminus X(\mathbb{R})$ is not connected. Otherwise the curve is said to be *nonseparating* (type II in Klein's classification).

The *topological type* is the set $(g, k, \varepsilon)$, where $g = g(X)$ is the genus of the curve, that is, the genus of the surface $X(\mathbb{C})$; $k = k(X)$ is the number of connected components of the set of real points $X(\mathbb{R})$; and

$$\varepsilon = \varepsilon(X) = \begin{cases} 0 & \text{if } X \text{ is nonseparating,} \\ 1 & \text{if } X \text{ is separating.} \end{cases}$$

In what follows we shall extensively use the fact that each Riemann surface $P$ is biholomorphically equivalent to a surface $H/\Gamma$, where $H$ is either the Riemann sphere $\overline{\mathbb{C}}$, or the complex plane $\mathbb{C}$, or the upper half-plane $\Lambda = \{z \in \mathbb{C} \mid \operatorname{Im} z > 0\}$, and $\Gamma$ is a discrete group acting without fixed points. The standard metric of constant curvature on $H$ induces a metric of constant curvature on $P = H/\Gamma$.

Let us give two examples of real algebraic curves.

EXAMPLE 1.1. Let $P$ be a surface of genus $\tilde{g}$ with $k$ holes. Equip $P$ with a Riemann surface structure $P^+$. Consider an atlas of holomorphic charts

$$\{(U_i, z_i)\}, \ P^+ = \bigcup U_i, \quad z_i \colon U_i \to \mathbb{C}.$$

The atlas $\{(U_i, \bar{z}_i)\}$ endows $P$ with another Riemann surface structure $P^-$. The natural mapping $\alpha\colon P^+ \to P \to P^-$ is antiholomorphic. The complex structures on $P^+$ and $P^-$ induce metrics of constant curvature on these surfaces, and $\alpha$ is an isometry with respect to these metrics. Let us encircle each hole in $P^+$ by a geodesic. The geodesics bound a compact surface $\tilde{P}^+ \subset P^+$ with the boundary $\varphi\tilde{P}^+$. Set $\tilde{P}^- = \alpha\tilde{P}^+$.

Now let us identify the boundaries $\varphi\tilde{P}^+$ and $\varphi\tilde{P}^-$ by means of $\alpha$. As a result, we obtain a compact Riemann surface $P_{\tilde{g},k}$ of genus $2\tilde{g} + k - 1$ and such that $\alpha$ induces an antiholomorphic involution $\tau_{\tilde{g},k}\colon P_{\tilde{g},k} \to P_{\tilde{g},k}$. Therefore, $X_{\tilde{g},k} = (P_{\tilde{g},k}, \tau_{\tilde{g},k})$ is a real algebraic curve and $X_{\tilde{g},k}(\mathbb{R}) = \varphi\tilde{P}^+ = \varphi\tilde{P}^-$. Hence $X_{\tilde{g},m}$ is a real algebraic curve of type $(2\tilde{g} + k - 1, k, 1)$.

EXAMPLE 1.2. Repeating the construction of Example 1.1, consider once more two Riemann surfaces with boundary $\tilde{P}^+, \tilde{P}^-$ and the antiholomorphic mapping $\alpha\colon \tilde{P}^+ \to \tilde{P}^-$. The boundary $\varphi\tilde{P}^+$ consists of some contours $c_1, \ldots, c_k$. Consider isometries $\alpha_i\colon c_i \to c_i$ without fixed points and such that $\alpha_i^2 = 1$. Now take $m$, $0 \leqslant m < k$. For $i \leqslant m$, we identify the contours $c_i$ and $\alpha c_i$ by means of $\alpha$. For $i > m$, we identify the contours $c_i$ and $\alpha c_i$ by means of $\alpha\alpha_i$. As a result, we obtain another real curve $Y_{\tilde{g},k}^m = (P_{\tilde{g},k}^m, \tau_{\tilde{g},k}^m)$ of the same genus. However, in this case $Y_{\tilde{g},k}^m(\mathbb{R}) = \bigcup_{i=1}^m c_i$, whence $Y_{\tilde{g},k}^m$ is a curve of topological type $(2\tilde{g} + k - 1, m, 0)$.

**2.** Two real curves $(P_1, \tau_1)$ and $(P_2, \tau_2)$ are said to be *topologically equivalent* if there is a homeomorphism $\phi\colon P_1 \to P_2$ such that $\tau_2\phi = \phi\tau_1$.

Our next goal is to prove that each real algebraic curve is topologically equivalent either to a curve in Example 1.1, or to a curve in Example 1.2.

LEMMA 1.1. *The set $X(\mathbb{R})$ of real points of a real algebraic curve $X = (P, \tau)$ decomposes into disjoint simple smooth contours (called ovals).*

Proof. The complex structure on $P$ induces on it a metric of constant curvature, and $\tau$ is an isometry with respect to this metric. For a point $x \in X(\mathbb{R})$, the involution $d\tau_x\colon T_x \to T_x$ of the tangent plane is a reflection in some line $v \in T_x$. Denote by $\ell \subset P$ the geodesic passing through $x$ in the direction $v$. All its points are fixed points of $\tau$, and these are the only fixed points in a small neighborhood of $x$. Hence each point $x \in X(\mathbb{R})$ belongs to exactly one geodesic $\ell \subset X(\mathbb{R})$, and this geodesic has no self-intersections. Since $P$ is compact, each such geodesic is a smooth closed contour. $\qquad\square$

THEOREM 1.1. *Let $(P, \tau)$ be a real curve of type $(g, k, 1)$. Then $1 \leqslant k \leqslant g + 1$, $k \equiv g + 1 \pmod 2$ and $(P, \tau)$ is topologically equivalent to a curve $(P_{\tilde{g},k}, \tau_{\tilde{g},k})$ in Example 1.1, where $\tilde{g} = \frac{1}{2}(g + 1 - k)$.*

Proof. By Lemma 1.1, the set $P \setminus P^\tau$ consists of two connected components $P_1$ and $P_2$, both of genus $\tilde{g}$ and with $k$ holes. Hence $g = 2\tilde{g} + k - 1$ and, therefore, $k \leqslant g + 1$, $k \equiv g + 1 \pmod 2$. Consider a homeomorphism

$\phi_1 \colon (P_1 \cup P^\tau) \to \tilde{P}^+$ and set

$$\phi(x) = \begin{cases} \phi_1(x) & \text{if } x \in P_1 \cup P^\tau, \\ \tau_{\tilde{g},k}\phi_1\tau(x) & \text{if } x \in P_2. \end{cases}$$

It is easy to see that $\phi$ realizes the desired topological equivalence.          □

**3.** Now we turn to nonseparating curves. Throughout this section, $Q$ is a Riemann surface of genus $g$ with $n$ holes and $\beta\colon Q \to Q$ is an antiholomorphic involution without fixed points.

A simple closed contour $a \subset Q$ is said to be *invariant* if $\beta a = a$.

A system of pairwise disjoint invariant contours $A = (a_1, \dots, a_m)$ is said to be *complete* if the set $Q \setminus A$ is disconnected. Obviously, if $A$ is complete, then $Q \setminus A$ consists of two surfaces $Q', Q''$, each of genus $\frac{1}{2}(g - m + 1)$ with $m + \frac{1}{2}n$ holes, and $\beta Q' = Q''$.

LEMMA 1.2. (a) *There is at least one invariant contour $a \subset Q$; (b) if $g > 0$, then there is an invariant contour $b \subset Q$ such that $Q \setminus b$ is connected; (c) there is a complete system consisting of $g + 1$ invariant contours; (d) if $A = (a_1, \dots, a_m) \subset Q$ is a complete system of invariant contours and $m > 2$, then there is a contour $b \subset Q$ such that $(a_1, \dots, a_{m-3}, b)$ also is a complete system of invariant contours.*

Proof. (a) Without loss of generality we may assume that $n > 2$. Consider on $Q$ the function $f(x) = \rho(x, \beta x)$, where $\rho$ is the distance with respect to the standard metric of constant curvature on $Q$. The function $f$ reaches its minimum $f(z) = c > 0$ at some point $z \in Q$. If $\ell$ is a minimal geodesic connecting $z$ and $\beta z$, then $a = \ell \cup \beta\ell$ is an invariant contour.

(b) Let $a \subset Q$ be the contour constructed above, and suppose $Q \setminus a$ is disconnected. Then $Q \setminus a = Q' \cup Q''$, where $Q', Q''$ are surfaces of positive genera and $\beta Q' = Q''$.

FIGURE 2.1.1

Connect points $x \in a$ and $\tau x$ by a non-self-intersecting line $\ell \subset Q'$ such that $Q' \setminus \ell$ is connected (see Fig. 2.1.1).

Then $b = \ell \cup \tau\ell$ is an invariant contour and $Q \setminus b$ is connected.

(c) Let $b$ be the contour constructed above. Then $Q \setminus b$ is a surface of genus $g - 1$, and if $g - 1 > 0$, then we can apply to it statement (b) once again. If $g = 0$, then we apply statement (a).

(d) The set $Q \setminus A$ decomposes into the surfaces $Q'$ and $Q''$ (see Fig. 2.1.2).

Let us add boundary contours to these surfaces. Two contours $a_i' \subset Q'$ and $a_i'' \subset Q''$ are associated to a contour $a_i \subset A$ in this process. Let $q_1, q_2, q_3$ be points of the contours $a_{m-2}, a_{m-1}, a_m$, and let $q_i'$ be the corresponding

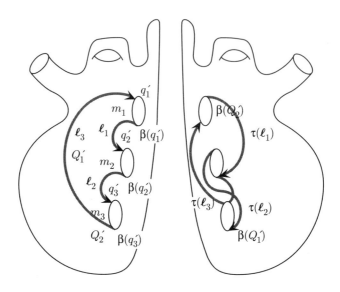

FIGURE 2.1.2

points of $a_{m-3+i}'$. Denote by $m_i$ one of the two arcs into which the points $q_i'$ and $\beta(q_i')$ split the contour $a_i'$. Connect the points $\beta(q_1')$ and $q_2'$ by a line $\ell_1 \subset Q'$ and the points $\beta(q_2')$ and $q_3'$ by a line $\ell_2 \subset Q'$ in such a way that $\ell_1$ and $\ell_2$ do not meet each other, are not self-intersecting, and do not intersect $\varphi Q'$. Connect the points $\beta(q_3')$ and $q_1'$ by a non-self-intersecting line $\ell_3$ homotopic to $(m_1 \ell_1 m_2 \ell_2 m_3)^{-1}$ and having no common points with it and with $\varphi Q'$ other than the ends (this can always be done, since the set $Q' \setminus (\ell_1 \cup \ell_2)$ is connected). The closed non-self-intersecting contour $\ell_3 m_1 \ell_1 m_2 \ell_2 m_3$ splits the surface $Q'$ into two parts $Q_1'$ and $Q_2'$. Now consider the invariant contour $b = \ell_1 \beta(\ell_2) \ell_3 \beta(\ell_1) \ell_2 \beta(\ell_3) \subset Q$. The surface $Q \setminus (b, a_1, \ldots, a_{m-3})$ decomposes into two surfaces $Q_1' \cup \beta(Q_2')$ and $Q_2' \cup \beta(Q_1')$. $\qquad\square$

THEOREM 1.2. *Let $(P, \tau)$ be a real algebraic curve of topological type $(g, m, 0)$. Then for any $m < k \leqslant g+1$, $k \equiv g+1 \pmod 2$ the curve $(P, \tau)$ is topologically equivalent to a curve $(P_{\tilde{g},k}^m, \tau_{\tilde{g},k}^m)$ in Example 1.2, where $\tilde{g} = \frac{1}{2}(g+1-k)$.*

Proof. Using Lemma 1.2, we find on $P \setminus P^\tau$ a complete set of contours $A = (a_{m+1}, \ldots, a_k)$, invariant with respect to $\tau$. The surface $P \setminus (P^\tau \cup A)$ decomposes into two surfaces $P_1$ and $P_2$, both of genus $\tilde{g}$ and with $k$ holes.

Now consider a homeomorphism $\phi_1 \colon (P_1 \cup P^\tau \cup A) \to \tilde{P}^+$ such that $\phi_1(P^\tau) = (c_1, \ldots, c_k)$. Set

$$\phi(x) = \begin{cases} \phi_1(x) & \text{if } x \in P_1 \cup P^\tau \cup A, \\ \tau_{\tilde{g},k}^m \phi_1 \tau(x) & \text{if } x \in P_2. \end{cases}$$

It is easy to see that $\phi$ establishes the desired topological equivalence.  $\square$

Examples 1.1, 1.2 and Theorems 1.1, 1.2 lead to the following

COROLLARY 1.1 ([93]). *Two real algebraic curves are topologically equivalent if and only if they have the same topological type. A set $(g, k, \varepsilon)$ is a topological type of some real algebraic curve if and only if either $\varepsilon = 1$, $1 \leqslant k \leqslant g+1$, $k \equiv g+1 \pmod 2$, or $\varepsilon = 0$, $0 \leqslant k \leqslant g$.*

**Remark.** The inequality $k \leqslant g+1$ for plane real algebraic curves was proved by Harnack [31], and it carries his name.

## 2. Moduli of real algebraic curves

**1.**   In the sequel we consider only the case $g > 1$. The case $g \leqslant 1$ is much simpler, but it requires different approaches.

Real algebraic curves of genus $g > 1$ are uniformized by discrete isometry groups of the metric $\frac{|dz|}{\operatorname{Im} z}$ on the Lobachevsky plane $\Lambda = \{z \in \mathbb{C} \,|\, \operatorname{Im} z > 0\}$. The complete group of isometries $\widetilde{\operatorname{Aut}}(\Lambda)$ consists of holomorphic ($\operatorname{Aut}(\Lambda)$) and antiholomorphic isometries.

Discrete subgroups $\Gamma \subset \widetilde{\operatorname{Aut}}(\Lambda)$ are called *non-Euclidean crystallographic groups* (NEC-groups) [44]. We shall need only NEC-groups $\tilde{\Gamma}$ such that $\Gamma = \tilde{\Gamma} \cap \operatorname{Aut}(\Lambda)$ is a Fuchsian group consisting of hyperbolic automorphisms, $\Gamma \neq \tilde{\Gamma}$, and $P = \Lambda/\Gamma$ is a compact surface. We call such Fuchsian groups $\tilde{\Gamma}$ *real Fuchsian groups*. For a real Fuchsian group, $\tilde{\Gamma} \setminus \Gamma$ induces an antiholomorphic involution $\tau = \Phi(\tilde{\Gamma} \setminus \Gamma)\Phi^{-1} \colon P \to P$ (where $\Phi \colon \Lambda \to P$ is the natural projection). Hence, a real Fuchsian group $\tilde{\Gamma}$ determines a real algebraic curve $(P, \tau) = [\tilde{\Gamma}]$.

LEMMA 2.1. *Each real algebraic curve is determined by some real Fuchsian group.*

Proof. Let $\Gamma \subset \operatorname{Aut}(\Lambda)$ be the Fuchsian group uniformizing a given Riemann surface $P$ and let $\Phi \colon \Lambda \to P$ be the natural projection (see, e.g., Section 2 of Chapter 1). Since $\Lambda$ is simply connected, there is an element $\sigma \in \widetilde{\operatorname{Aut}}(\Lambda) \setminus \operatorname{Aut}(\Lambda)$ such that $\Phi\sigma = \tau\Phi$. Let $\tilde{\Gamma}$ be the group generated by $\sigma$ and $\Gamma$. Then $(P, \tau) = [\tilde{\Gamma}]$.  $\square$

**2.**   Let $M_{g,k,\varepsilon}$ denote the moduli space of real algebraic curves of type $(g, k, \varepsilon)$. Our immediate goal is to construct a natural mapping $\Psi_{\tilde{g},k}^k \colon \tilde{T}_{\tilde{g},k} \to M_{g,k,1}$, where $\tilde{g} = \frac{1}{2}(g+1-k)$.

Let $n = \tilde{g} + k$, $\psi \in \tilde{T}_{\tilde{g},k}$ and let $\{A_i, B_i \ (i = 1, \ldots, \tilde{g}), C_i \ (i = 1, \ldots, k)\} = \{\psi(a_i), \psi(b_i) \ (i = 1, \ldots, \tilde{g}), \psi(c_i) \ (i = \tilde{g}+1, \ldots, n)\}$. We denote by $\overline{C}_i \in$

$\widetilde{\mathrm{Aut}}(\Lambda) \setminus \mathrm{Aut}(\Lambda)$ the reflection (in the sense of Lobachevsky geometry) with respect to the geodesic $\ell(C_i)$. Let $\Gamma_\psi = \psi(\gamma_{\tilde{g},n})$ and let $\Gamma_\psi^k$ be the group generated by $\Gamma_\psi$ and $\overline{C}_1, \dots, \overline{C}_k$.

LEMMA 2.2. *The group $\Gamma_\psi^k$ is a real Fuchsian group and $[\Gamma_\psi^k] \in M_{g,k,1}$.*

Proof. Let $\{\tilde{a}_i, \tilde{b}_i \ (i = 1, \dots, g)\}$ be generators of $\gamma_{g,0}$, subject to the defining relation $\prod_{i=1}^g [\tilde{a}_i, \tilde{b}_i] = 1$. We set

$$\tilde{\psi}(\tilde{a}_i) = \overline{C}_n B_{\tilde{g}+1-i} \overline{C}_n, \quad \tilde{\psi}(\tilde{b}_i) = \overline{C}_n A_{\tilde{g}+1-i} \overline{C}_n \quad (i = 1, \dots, \tilde{g}),$$
$$\tilde{\psi}(\tilde{a}_i) = A_{i-\tilde{g}}, \quad\quad\quad \tilde{\psi}(\tilde{b}_i) = B_{i-\tilde{g}} \quad\quad\quad (i = \tilde{g}+1, \dots, 2\tilde{g}),$$
$$\tilde{\psi}(\tilde{a}_i) = W_i C_i W_i, \quad\quad \tilde{\psi}(\tilde{b}_i) = W_i D_i W_i^{-1} \quad\quad (i = 2\tilde{g}+1, \dots, 2\tilde{g}+k),$$

where $D_i = \overline{C}_n \overline{C}_i$, $W_i = \prod_{j=i-1}^1 D_j C_j D_j^{-1}$ (see Fig. 2.2.1).

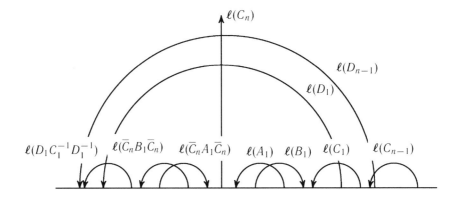

FIGURE 2.2.1

We have

$$\prod_{i=1}^g [\tilde{\psi}(\tilde{a}_i), \tilde{\psi}(\tilde{b}_i)] = \overline{C}_n \prod_{i=\tilde{g}}^1 [B_i A_i] \overline{C}_n \prod_{i=1}^{\tilde{g}} [A_i B_i] \prod_{i=1}^k C_i \prod_{i=k}^1 \overline{C}_n C_i^{-1} \overline{C}_n = 1$$

because of $\prod_{i=1}^{\tilde{g}} [A_i B_i] \prod_{i=1}^k C_i = 1$. Moreover,

$$(\tilde{\psi}(\tilde{a}_1), \tilde{\psi}(\tilde{b}_1 \tilde{a}_1^{-1} \tilde{b}_1^{-1}), \dots, \tilde{\psi}(\tilde{a}_g), \tilde{\psi}(\tilde{b}_g \tilde{a}_g^{-1} \tilde{b}_g^{-1}))$$

is a sequential set of type $(0, 2g)$ (see Fig. 2.2.1). Hence $\tilde{\psi} \in \tilde{T}_{g,0}$ and, therefore, $P = \Lambda/\tilde{\psi}(\gamma_{g,0}) \in M_{g,0}$. The group $\Gamma_\psi^k$ is generated by the group $\tilde{\psi}(\gamma_{g,0})$ and the involutions $\overline{C}_i$, and $\overline{C}_i \tilde{\psi}(\gamma_{g,0}) \overline{C}_i = \tilde{\psi}(\gamma_{g,0})$. Therefore, $\Gamma_\psi^k$ is a real Fuchsian group and the images $\ell(C_i)$ form the ovals of the curve $[\Gamma_\psi^k]$. By the construction, these contours form the boundary of a surface of genus $\tilde{g}$. $\square$

Thus the correspondence $\psi \mapsto [\Gamma_\psi^k]$ determines a mapping $\Psi_{\tilde{g},k}^k : \tilde{T}_{\tilde{g},k} \to M_{g,k,1}$.

LEMMA 2.3. $\Psi^k_{\tilde{g},k}(\tilde{T}_{\tilde{g},k}) = M_{g,k,1}$.

Proof. Let $(P,\tau) \in M_{g,k,1}$. By Lemma 2.1, $(P,\tau) = [\tilde{\Gamma}]$ for some real Fuchsian group $\tilde{\Gamma}$. Let $\Gamma \in \tilde{\Gamma} \cap \operatorname{Aut}(\Lambda)$, let $\Phi\colon \Lambda \to \Lambda/\Gamma = P$ be the natural projection, $\Phi(q) = p$, and let $\Phi_q\colon \gamma \to \pi_1(P,p)$ be an isomorphism taking an element $h \in \Gamma$ to the image $\Phi(\ell)$ of the segment $\ell = [q, hq] \subset \Lambda$. The ovals $P^\tau$ split the surface $P$ into two surfaces $P_1$ and $P_2$. Let $p \in P_1$ and let $v = \{a_i, b_i \ (i = 1, \ldots, \tilde{g}), c_i \ (i = \tilde{g}+1, \ldots, n)\}$ be a standard system of generators of $\pi(P_1, p)$ in the sense of Section 2, Chapter 1. By Theorem 2.1 of Chapter 1, $V = \Phi_q^{-1}(v)$ is a sequential set of type $(\tilde{g}, k)$, i.e., $V = \psi(v)$, where $\psi \in \tilde{T}_{\tilde{g},k}$. But this means that $[\Gamma^k_\psi] = (P,\tau)$. $\qquad \square$

According to Section 4 of Chapter 1,

$$T_{\tilde{g},k} = \tilde{T}_{\tilde{g},k}/\operatorname{Aut}(\Lambda) \cong \mathbb{R}^{6\tilde{g}+3k-6} = \mathbb{R}^{3g-3}.$$

For obvious reasons, the mapping

$$\Psi^k_{\tilde{g},k}\colon \tilde{T}_{\tilde{g},k} \to M_{g,k,1}$$

induces the mapping

$$\Psi^k_{\tilde{g},k}\colon T_{\tilde{g},k} \to M_{g,k,1}.$$

We shall also make use of the mapping

$$\alpha\colon T_{\tilde{g},k} \to T_{\tilde{g},k}$$

defined by its values

$$\alpha\psi(a_i) = \beta\psi(b_{\tilde{g}+1-i})\beta, \quad \alpha\psi(b_i) = \beta\psi(a^{-1}_{\tilde{g}+1-i})\beta \quad (i = 1, \ldots, \tilde{g}),$$
$$\alpha\psi(c_i) = w\beta\psi(c^{-1}_{\tilde{g}+k+1-i})\beta w^{-1} \qquad (i = \tilde{g}+1, \ldots, \tilde{g}+k),$$

where $\beta(z) = -\bar{z}$ and $w = \alpha\psi\left(\prod_{i=1}^{\tilde{g}}[a_i, b_i]\right)$. Let $\operatorname{Mod}^k_{\tilde{g},k}$ be the group of automorphisms of $T_{\tilde{g},k}$ generated by $\operatorname{Mod}_{\tilde{g},k}$ and $\alpha$. Then $\operatorname{ind}(\operatorname{Mod}_{\tilde{g},k} : \operatorname{Mod}^k_{\tilde{g},k}) = 2$. Moreover, it is obvious from the construction that $[\Gamma^k_\psi] = [\Gamma^k_{\psi'}]$ if and only if $\psi' = \gamma\psi$, where $\gamma \in \operatorname{Mod}^k_{\tilde{g},k}$. Hence, Lemmas 2.2 and 2.3 imply

THEOREM 2.1 ([53], [54], [55]). $M_{g,k,1} = T_{\tilde{g},k}/\operatorname{Mod}^k_{\tilde{g},k}$, and the group $\operatorname{Mod}^k_{\tilde{g},k}$ acts discretely.

**3.** Now we describe the space $M_{g,m,0}$. To this end, let us construct a mapping

$$\Psi^m_{\tilde{g},k}\colon \tilde{T}_{\tilde{g},k} \to M/T_{g,m,0},$$

where $m < k$, $k \equiv g+1 \pmod 2$ and $\tilde{g} = \frac{1}{2}(g+1-k)$.

Suppose as before that to each monomorphism $\psi \in \tilde{T}_{\tilde{g},k}$ a sequential set

$$\{A_i, B_i \ (i = 1, \ldots, \tilde{g}), C_i \ (i = 1, \ldots, k)\}$$
$$= \{\psi(a_i), \psi(b_i) \ (i = 1, \ldots, \tilde{g}), \psi(c_i) \ (i = \tilde{g}+1, \ldots, n)\}$$

is assigned. Set $\tilde{C}_i = \overline{C}_i \sqrt{C_i}$, where $\sqrt{C_i}$ is a hyperbolic automorphism such that $(\sqrt{C_i})^2 = C_i$. Let $\Gamma_\psi = \psi(\gamma_{\tilde{g},n})$ and let $\Gamma_{\psi,k}^m$ be the group generated by $\Gamma_\psi$, $\overline{C}_1, \ldots, \overline{C}_m$ and $\tilde{C}_{m+1}, \ldots, \tilde{C}_k$.

LEMMA 2.4. *The group $\Gamma_{\psi,k}^m$ is a real Fuchsian group and $[\Gamma_{\psi,k}^m] \in M_{g,m,0}$.*

Proof. The proof repeats almost literally that of Lemma 2.2. The only difference is that the images $\ell(C_i)$ form ovals if and only if $i \leqslant m$, and that is why the curve $[\Gamma_{\psi,k}^m]$ is nonseparating.                    $\square$

Hence the correspondence $\psi \mapsto [\Gamma_{\psi,k}^m]$ defines the mapping

$$\Psi_{\tilde{g},k}^m : \tilde{T}_{\tilde{g},k} \to T_{g,m,0}.$$

LEMMA 2.5. $\Psi_{\tilde{g},k}^m(\tilde{T}_{\tilde{g},k}) = M_{g,m,0}$.

Proof. Let $(P, \tau) \in M_{g,m,0}$. By Theorem 1.2, there is a set of invariant contours $A = (a_{m+1}, \ldots, a_k) \in P \setminus P^\tau$ such that $P \setminus (P^\tau \cup A)$ decomposes into two surfaces $P_1$ and $P_2$, both of genus $\tilde{g} = \frac{1}{2}(g + 1 - k)$. The remaining part of the proof repeats almost literally that of Lemma 2.3.          $\square$

THEOREM 2.2 ([**53**], [**54**], [**55**]). $M_{g,k,0} = T_{\tilde{g},k} / \operatorname{Mod}_{\tilde{g},k}^m$, *the group* $\operatorname{Mod}_{\tilde{g},k}^m$ *acts discretely and* $\operatorname{ind}(\operatorname{Mod}_{\tilde{g},k}^m \cap \operatorname{Mod}_{\tilde{g},k}^k : \operatorname{Mod}_{\tilde{g},k}^k) = \binom{k}{m}$.

Proof. The mapping $\Psi_{g,k}^m : \tilde{T}_{\tilde{g},k} \to M_{g,m,0}$ induces a mapping $\Psi_{g,k}^m : T_{\tilde{g},k} \to M_{g,m,0}$ in an obvious way. Let $\Psi_{\tilde{g},k}^m(\psi) = \Psi_{\tilde{g},k}^m(\psi')$, i.e., $(P, \tau) = [\Gamma_{\psi,k}^m] = [\Gamma_{\psi',k}^m] = (P', \tau')$. Consider the monomorphisms $\psi, \psi' \in T_{g,0}$. Then $\tilde{\psi}' = \tilde{\psi}\gamma$, where $\gamma$ belongs to the group $\operatorname{Mod}_{\tilde{g},k}^m$ generated by the group $\{\gamma \in \operatorname{Mod}_{g,0} \mid \gamma\tau = \tau\gamma\}$ and the element $\alpha$, and for any $\gamma \in \operatorname{Mod}_{\tilde{g},k}^m$ we have $\Psi_{\tilde{g},k}^m(\psi\gamma) = \Psi_{\tilde{g},k}^m(\psi)$. Now consider the subgroup $\operatorname{Mod}_{\tilde{g},k}^m \cap \operatorname{Mod}_{\tilde{g},k}^k$ consisting of those automorphisms in $\operatorname{Mod}_{\tilde{g},k}^k$ that preserve the sets $c_i$ $(i = 1, \ldots, m)$. It is easy to see that its index in $\operatorname{Mod}_{\tilde{g},k}^k$ is $\binom{k}{m}$.          $\square$

Comparing Theorems 2.1, 2.2 with Section 4 of Chapter 1 we obtain

COROLLARY 2.1 ([**53**], [**54**], [**55**]). *The moduli space of real algebraic curves of genus $g > 1$ decomposes into connected components $M_{g,k,\varepsilon}$, where $(g, k, \varepsilon)$ is an arbitrary topological type of a real algebraic curve. Each connected component is diffeomorphic to $\mathbb{R}^{3g-3} / \operatorname{Mod}_{g,k,\varepsilon}$, where $M_{g,k,\varepsilon}$ is a discrete group of automorphisms.*

**Remark.** The statement about the topological structure of the connected components of the space of real algebraic curves in Corollary 2.1 appeared first in [**24**]. The proof in [**24**] proceeded in the framework of quasiconformal mappings and was based on a theorem in [**42**], which was afterwards discovered to be incorrect. A correct proof based on the theory of quasiconformal mappings can be found in [**90**].

## 3. Arf functions on real algebraic curves

**1.** In the study of spinor bundles and super-Riemann surfaces, the Arf functions play an important role (see Sections 7–13 of Chapter 1). To real algebraic curves, special Arf functions, which we start describing, are associated.

Let $P$ be a surface of genus $g = g(P)$ with $k$ holes. A basis $v = \{a_i, b_i \ (i = 1, \ldots, g), \ c_i \ (i = g+1, \ldots, g+k)\}$ of $H_1(P, \mathbb{Z}_2)$ (where $\mathbb{Z}_2 = \mathbb{Z}/2\mathbb{Z} = \{0, 1\}$) is said to be *standard* if the generators $c_i$ correspond to the holes in $P$ and $(a_i, a_j) = (b_i, b_j) = 0$, $(a_i, b_j) = \delta_{ij}$, where $(.\,,\,.) \in \mathbb{Z}_2$ is the intersection number in $H_1(P, \mathbb{Z}_2)$.

An *Arf function* on $P$ is a function $\omega \colon H_1(P, \mathbb{Z}_2) \to \mathbb{Z}_2$ such that $\omega(a + b) = \omega(a) + \omega(b) + (a, b)$. We say that an Arf function $\omega$ is *even* and set $\delta = \delta(P, \omega) = 0$ if there is a standard basis $v$ such that

$$\sum_{i=1}^{g} \omega(a_i)\omega(b_i) \equiv 0 \pmod 2.$$

Otherwise, we set $\delta = \delta(P, \omega) = 1$ and say that $\omega$ is *odd*. Denote by $k_\alpha = k_\alpha(P, \omega)$ $(\alpha = 0, 1)$ the number of elements $c_i$ in the standard basis such that $\omega(c_i) = \alpha$. The set $(g, \delta, k_\alpha)$ is called the *topological type* of the Arf function $\omega$.

According to Section 8 of Chapter 1, a set $(g, \delta, k_\alpha)$ is a topological type of some Arf function if and only if $k_1 \equiv 0 \pmod 2$ and $\delta = 0$ for $k_1 > 0$. If this is the case, then there is a standard basis $v$ such that $\omega(a_i) = \omega(b_i) = 0$ for $i > 1$ and $\omega(a_1) = \omega(b_1) = \delta$.

Two Arf functions $\omega_1$ and $\omega_2$ on $P$ are said to be *topologically equivalent* if there is a homeomorphism $\psi \colon P \to P$ inducing an automorphism $\tilde{\psi} \colon H_1(P, \mathbb{Z}_2) \to H_1(P, \mathbb{Z}_2)$ such that $\omega_1 = \omega_2 \tilde{\psi}$.

By Section 8 of Chapter 1, two Arf functions are topologically equivalent if and only if they have the same topological type.

**2.** Let $(P, \tau)$ be a real algebraic curve. Below we denote a simple contour and its homology class in $H_1(P, \mathbb{Z}_2)$ by the same letter. The involution $H_1(P, \mathbb{Z}_2) \to H_1(P, \mathbb{Z}_2)$ induced by $\tau \colon P \to P$ will also be denoted by $\tau$.

An *Arf function on a real algebraic curve* $(P, \tau)$ (or, simply, a *real Arf function*) is an Arf function $\omega \colon H_1(P, \mathbb{Z}_2) \to \mathbb{Z}_2$ such that $\omega\tau = \omega$.

LEMMA 3.1. *Let $(P, \tau)$ be a real curve, let $c_1, c_2 \subset P$ be simple closed curves such that $\tau(c_i) = c_i$, $c_i \cap P^\tau = \varnothing$, $c_1 \cap c_2 = \varnothing$, and let $\omega$ be an arbitrary Arf function on $(P, \tau)$. Then $\omega(c_1) = \omega(c_2)$.*

Proof. By Theorem 1.2, there is a set of pairwise disjoint simple contours $c_3, \ldots, c_r \subset P \setminus (c_1 \cup c_2)$ such that $\tau(c_i) = c_i$ and $P \setminus \bigcup_{i=1}^{r} c_i$ decomposes into surfaces $P_1$ and $P_2$, where $\tau P_1 = P_2$. Connect the contours $c_1$ and $c_2$ by a non-self-intersecting line $\ell \subset P_1$. Let $d$ be a simple closed contour of the form

$$d = \ell \cup f_1 \cup \tau\ell \cup f_2,$$

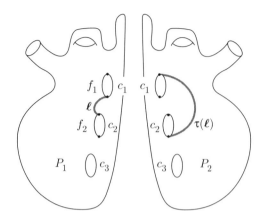

FIGURE 2.3.1

where $f_i \subset c_i$ is the segment connecting the points $\ell \cap c_i$ and $\tau\ell \cap c_i$ (see Fig. 2.3.1). Then $\tau(d) = d + c_1 + c_2$, whence,

$$\omega(d) = \omega(d) + \omega(c_1) + \omega(c_2). \qquad \square$$

An Arf function $\omega$ on $(P, \tau)$ is said to be *special* if there is a simple closed contour $c$ such that $\tau(c) = c$, $c \cap P^\tau = \varnothing$ and $\omega(c) = 0$.

LEMMA 3.2. *If $P^\tau \neq \varnothing$, then any real Arf function on $(P, \tau)$ is nonspecial.*

Proof. Let $c \subset P$ be a simple contour such that $\tau(c) = c$ and $c \cap P^\tau = \varnothing$. Let $c' \subset P^\tau$ be one of the ovals of the real curve $(P, \tau)$. Then, by Theorem 1.2, there is a set of pairwise disjoint simple contours $c_1, \ldots, c_r \in P \setminus (c \cup c')$ such that $\tau(c_i) = c_i$ and $P \setminus \left(c \cup c' \cup \bigcup_{i=1}^{r} c_i\right)$ decomposes into surfaces $P_1, P_2$, $\tau P_1 = P_2$. Connect the contours $c$ and $c'$ by a non-self-intersecting line $\ell \subset P_1$. Let $d$ be a simple closed contour of the form $d = \ell \cup f \cup \tau\ell$, where $f \subset c$. Then $\tau(d) = d + c$ and, therefore, $\omega(d) = \omega(d + c) = \omega(d) + \omega(c) + 1$. $\qquad \square$

LEMMA 3.3. *A special real Arf function vanishes on all invariant contours.*

Proof. Let $\omega$ be a special Arf function on a real algebraic curve $(P, \tau)$. Suppose there is a contour $c \in P$ such that $\tau c = c$ and $\omega(c) = 1$. According to Lemmas 1.2 and 3.2, there is a complete system of contours $c, c_1, \ldots, c_g$ cutting $P$ into spheres with holes $P_1$ and $P_2$. Connect the contour $c_i$ with the contour $c$ by a segment $\ell_i \subset P_1$ and set

$$d_i = \ell_i \cup \tau\ell_i \cup r_i \cup r,$$

where $r_i \subset c_i$ (respectively, $r \subset c$) are arcs connecting the points $p_i = \ell_i \cap c_i$ and $\tau p_i$ (respectively, the points $p = \ell_i \cap c$ and $\tau p$). Consider a disk $D_1 \subset P_1$. Let us identify the boundary contours of the surface $P \setminus (D_1 \cup \tau D_1)$ by means of $\tau$. On the surface $\tilde{P}$ thus obtained the involution $\tau$ induces an involution $\tilde{\tau}$ with the oval $\tilde{c} = \partial D_1$. Connect $\tilde{c}$ with $c$ by a segment $\tilde{\ell} \subset P_1$ and set $\tilde{d} = \tilde{\ell} \cup \tau\tilde{\ell} \cup \tilde{r}$, where $\tilde{r} \subset c$ is the arc connecting the points $\tilde{p} = \tilde{\ell} \cap c$ and $\tau\tilde{p}$. The contours $\{c_i, d_i \ (i = 1, \ldots, g), \tilde{c}, \tilde{d}\}$ form a basis in $H_1(\tilde{P}, \mathbb{Z}_2)$. Consider the Arf function $\tilde{\omega}$ on $\tilde{P}$ such that $\tilde{\omega}(c_i) = \omega(c_i)$, $\tilde{\omega}(d_i) = \omega(d_i)$, $\tilde{\omega}(\tilde{c}) = \tilde{\omega}(\tilde{d}) = 0$. Then $\tilde{\omega}(c) = \sum_{i=1}^{g} \tilde{\omega}(c_i) = \sum_{i=1}^{g} \omega(c_i) = \omega(c)$, $\tilde{\omega}(\tau\tilde{d}) = \tilde{\omega}(\tilde{d} + c) = \tilde{\omega}(\tilde{d}) + \tilde{\omega}(c) + 1 = \tilde{\omega}(\tilde{d})$ and, therefore, $\tilde{\omega}$ is a real Arf function. By Lemma 3.2, this implies that $\tilde{\omega}$ equals 1 on all contours $c'$ in $P \setminus \tilde{c}$ such that $\tau c' = c'$. But the Arf functions $\omega$ and $\tilde{\omega}$ coincide on these contours, whence $\omega$ is nonspecial. The contradiction thus obtained proves that $\omega(c) = 0$. $\qquad\square$

THEOREM 3.1 ([**76**]). *A special Arf function on a real curve* $(P, \tau)$ *of type* $(g, k, \varepsilon)$ *exists if and only if* $k = \varepsilon = 0$. *If this assumption is satisfied, then there are* $2^g$ *Arf functions, all of them even.*

Proof. The assertion $k = \varepsilon = 0$ for special Arf functions follows from Lemma 3.2. Now suppose $k = \varepsilon = 0$. Consider a standard basis $\{c_i, d_i \ (i = 1, \ldots, g)\} \subset H_1(P, \mathbb{Z}_2)$ such that $\tau c_i = c_i$, $\tau d_i = d_i + c_i + \sum_{i=1}^{g} c_i$, constructed in the course of the proof of Lemma 3.3. We set $w(c_i) = 0$ for all $i$, assign to $\omega(d_i) \ (i = 1, \ldots, g)$ arbitrary values in $\mathbb{Z}_2$, and extend $\omega$ to $H_1(P, \mathbb{Z}_2)$ by setting $\omega(a + b) = \omega(a) + \omega(b) + (a, b)$. Then $\omega(\tau d_i) = \omega(d_i)$, and therefore, $\omega$ is a special even Arf function. By Lemma 3.3, this construction produces all Arf functions on $(P, \tau)$. $\qquad\square$

**3.**      The *topological type* of an Arf function $\omega$ on a real curve $(P, \tau)$ of type $(g, k, 0)$ is the set $(g, \delta, k_\alpha)$, where $\delta = \delta(P, \omega)$ and $k_\alpha \ (\alpha = 0, 1)$ is the number of ovals $c_i \in P^\tau$ such that $\omega(c_i) = \alpha$.

THEOREM 3.2 ([**76**]). *A set* $(g, \delta, k_\alpha)$ *is the topological type of some nonspecial Arf function on a real curve of type* $(g, k, 0)$ *if and only if* $k = k_0 + k_1 \leqslant g$ *and* $k_0 = g + 1 \ (\mathrm{mod}\ 2)$. *If these requirements are satisfied, then the number of such functions equals* $\binom{k}{k_0} \cdot 2^{g-1}$.

Proof. Let $(P, \tau)$ be a real curve of type $(g, k, 0)$. By Theorem 1.2, there is a set of pairwise disjoint smooth contours $(c_1, \ldots, c_{g+1})$ such that $P^\tau = \bigcup_{i=1}^{k} c_i$ and $\tau(c_i) = c_i$. These contours cut $P$ into two spheres $P_1$ and $P_2$, each with $g + 1$ holes, and $\tau P_1 = P_2$. Let $\omega$ be a nonspecial Arf function on $(P, \tau)$. Then, by Section 8 of Chapter 1, the Arf function $\omega|_{P_1}$ takes value 1 on evenly many holes. Therefore, if $\omega$ is nonsingular, then $k_1 + (g + 1 - k) \equiv 0 \ (\mathrm{mod}\ 2)$, i.e., $k_0 \equiv g + 1 \ (\mathrm{mod}\ 2)$.

Now let $(g, \delta, k_\alpha)$ be an arbitrary set such that $k_0 + k_1 \leqslant g$ and $k_0 \equiv g + 1$ (mod 2). Let us connect the contours $c_i$ and $c_{g+1}$ by a segment $\ell_i \subset P_i$. Consider the simple contour $d_i = \ell_i \cup \tau\ell_i \cup r_i \cup r_{g+1}$, where $r_j \subset c_j$. Then $\tau(d_i) = d_i + c_{g+1} + \alpha_i c_i$, where $\alpha_i = 0$ for $i \leqslant k$ and $\alpha_i = 1$ for $i > k$.

Now let us set $\omega(c_i) = 0$ for arbitrary $k_0$ contours among $c_1, \ldots, c_k$, and $\omega(c_i) = 1$ for the rest of them. Since $k_0 \equiv g + 1 \ (\mathrm{mod}\ 2)$, the latter set is nonempty. Let $c_r$ be one of the contours in this set, that is, $\omega(c_r) = 1$. We choose arbitrary values $\omega(d_i)$ for $i \neq r$ and set

$$\omega(d_r) = \delta - \sum_{i \neq r} \omega(c_i)\omega(d_i).$$

Now extend $\omega$ to the entire space $H_1(P, \mathbb{Z}_2)$ by setting

$$\omega(a + b) = \omega(a) + \omega(b) + (a, b).$$

It is easy to see that this construction produces all real nonspecial Arf functions of type $(g, \delta, k_\alpha)$. $\qquad\square$

**4.** On separating curves, the Arf functions (which are automatically nonspecial) possess additional topological invariants.

Let $(P, \tau)$ be a separating real curve and let $P_1 \cup P_2 = P \setminus P^\tau$. Connect the ovals $c_i, c_j \in P^\tau$ by a segment $\ell_{ij} \subset P_1$. Consider the contours $d_{ij} = \ell_{ij} \cup \tau\ell_{ij}$. We call two ovals $c_i$ and $c_j$ $\omega$-similar on $(P, \tau)$, provided $\omega(d_{ij}) = 0$.

THEOREM 3.3. *The $\omega$-similarity is a well-defined equivalence relation, and it splits the ovals into at most two equivalence classes.*

Proof. Let $\tilde{\ell}_{ij} \subset P_1$ be another segment connecting $c_i$ and $c_j$, let $\tilde{d}_{ij} = \tilde{\ell}_{ij} \cup \tau\tilde{\ell}_{ij}$ and let $b \subset P_1 \cup P^\tau$ be the closed contour constructed of $\ell_{ij}, \tilde{\ell}_{ij}$, and parts of the ovals $c_i$ and $c_j$. Then $\omega(d_{ij} + \tilde{d}_{ij}) = \omega(b + \tau b) = 2\omega(b) = 0$, whence $\omega(d_{ij}) = \omega(d_{ij} + \tilde{d}_{ij}) + \omega(\tilde{d}_{ij}) = \omega(\tilde{d}_{ij})$. Therefore, the $\omega$-similarity

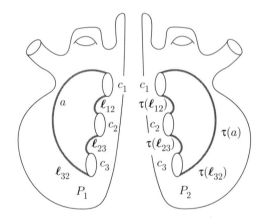

FIGURE 2.3.2

property does not depend on the choice of the segment $\ell_{ij}$. Now let $a \subset P_1 \cup P^\tau$ be the closed contour consisting of the segments $\ell_{ij}, \ell_{jk}, \ell_{ki}$, and the parts of the ovals $c_i, c_j, c_k$ (see Fig. 2.3.2). Then $\omega(d_{ij} + d_{jk} + d_{ki}) = \omega(a + \tau a) = 2\omega(a) = 0$, whence $\omega(d_{ij}) = \omega(d_{ik}) + \omega(d_{kj})$. Thus, if $c_i$ is (respectively, is not) $\omega$-similar to $c_k$ and $c_k$ is (respectively, is not) $\omega$-similar to $c_j$, then $c_i$ is similar to $c_j$. $\qquad\square$

Choose an oval $c \in P^\tau$. Let $B_c$ be the set of ovals $\omega$-similar to $c$. Denote by $k_\alpha^0 = k_\alpha^0(P, \tau, \omega)$ (respectively, $k_\alpha^1 = k_\alpha^1(P, \tau, \omega)$) the number of ovals $c_i$

in $B_c$ (respectively, in $P^\tau \setminus B_c$) such that $\omega(c_i) = \alpha$. The set of numbers $k_\alpha^\gamma$ $(\alpha, \gamma \in \{0, 1\})$ is well defined up to the global transformation $k_\alpha^\gamma \mapsto k_\alpha^{1-\gamma}$ related to the choice of the contour $c$.

The *topological type of an Arf function* $\omega$ on a real curve $(P, \tau)$ of type $(g, k, 1)$ is the set $(g, \tilde{\delta}, k_\alpha^\gamma)$, where $k_\alpha^\gamma = k_\alpha^\gamma(P, \tau, \omega)$, $\tilde{\delta} = \delta(P_1, \omega|_{P_1})$ and $P_1 \cup P_2 = P \setminus P^\tau$.

THEOREM 3.4 ([**76**]). *A set $(g, \tilde{\delta}, k_\alpha^\gamma)$ is the topological type of an Arf function on a real curve $(P, \tau)$ of type $(g, k, 1)$ if and only if $(\tilde{g}, \tilde{\delta}, k_\alpha^0 + k_\alpha^1)$ is the topological type of some Arf function on a surface of genus $\tilde{g} = \frac{1}{2}(g + 1 - k)$ with $k$ holes. If this is the case, then the number of such Arf functions is*

$$\binom{k}{k_0} \cdot \binom{k_0}{k_0^0} \cdot \binom{k_1}{k_1^0} \cdot 2^{\tilde{g}-2} \cdot (2^{\tilde{g}} + m),$$

*where $m = 2^{\tilde{g}}$ if $k_1 > 0$, $m = 1$ if $\tilde{\delta} = 0$, and $m = -1$ if $k_1 = 0$ and $\tilde{\delta} = 1$. The parity of the Arf function coincides with that of $k_1^0$.*

Proof. If $(g, \tilde{\delta}, k_\alpha^\gamma)$ is the topological type of an Arf function on a real curve $(P, \tau)$ of type $(g, k, 1)$, then the set $(\tilde{g}, \tilde{\delta}, k_\alpha^0 + k_\alpha^1)$ is the topological type of the Arf function $\omega|_{P_1} \colon H_1(P_1, \mathbb{Z}_2) \to \mathbb{Z}_2$, where $P \setminus P^\tau = P_1 \cup P_2$. Now suppose $(P, \tau)$ is a real curve of type $(g, k, 1)$, $P \setminus P^\tau = P_1 \cup P_2$, $\tilde{\omega} \colon H_1(P_1, \mathbb{Z}_2) \to \mathbb{Z}_2$ is an Arf function on $P_1$ of type $(\tilde{g}, \tilde{\delta}, k_\alpha^0 + k_\alpha^1)$ and $v = \{a_i, b_i \ (i = 1, \ldots, \tilde{g}), c_j \ (j = \tilde{g} + 1, \ldots, \tilde{g} + k)\} \subset H_1(P_1, \mathbb{Z}_2)$ is a standard basis. Group the ovals $c_i$ in groups $A_0^0, A_0^1, A_1^0, A_1^1$, where $A_\alpha^\gamma$ contains $k_\alpha^\gamma$ contours, in an arbitrary way. Connect the ovals $c_i$ and $c_k$ by segments $\ell_i \subset P_1$ and set $d_i = \ell_i \cup \tau\ell_i$. We assume that $\omega(c_i) = \alpha$ if $c_i \in A_\alpha^0 \cup A_\alpha^1$, and $\omega(d_i) = 0$ if $c_i$ and $c_k$ belong to the same subset $A_0^\alpha \cup A_1^\alpha$. Otherwise we set $\omega(d_i) = 1$. Finally, set

$$\omega(\tau a_i) = \omega(a_i), \quad \omega(\tau b_i) = \omega(b_i) \quad (i = 1, \ldots, \tilde{g}).$$

The equality

$$\omega(a + b) = \omega(a) + \omega(b) + (a, b)$$

determines a unique extension of $\omega$ to an Arf function on $(P, \tau)$. It is easy to see that $\omega$ has the type $(g, \tilde{\delta}, k_\alpha^\gamma)$, and the construction above produces all Arf functions of this type. The Arf function $\omega$ is even if $k_1 = 0$, and for $k_1 > 0$ its parity coincides with the parity of the number of contours in $A_1^0$ (recall that the sum $k_1^0 + k_1^1$ is even). $\square$

## 4. Lifting real Fuchsian groups

**1.** Denote by

$$J \colon \mathrm{SL}(2, \mathbb{R}) \to \mathrm{PSL}(2, \mathbb{R}) = \mathrm{Aut}(\Lambda)$$

the natural projection. Let

$$\Gamma \subset \mathrm{Aut}(\Lambda)$$

be a Fuchsian group consisting of hyperbolic automorphisms. We call a subgroup $\Gamma^* \subset \mathrm{SL}(2, R)$ a *lifting* of $\Gamma$ if $J(\Gamma^*) = \Gamma$ and $J|_{\Gamma^*} \colon \Gamma^* \to \Gamma$ is an isomorphism.

According to Section 7 of Chapter 1, associated to a lifting $\Gamma^*$ is an Arf function

$$\omega_{\Gamma^*} \colon H_1(\Lambda/\Gamma, \mathbb{Z}_2) \to \mathbb{Z}_2.$$

It is defined as follows. Let $a' \in \Gamma$ and let $a \in H_1(\Lambda/\Gamma, \mathbb{Z}_2)$ be the image of $a'$ under the natural projection $P_r \colon \Gamma \to \pi_1(\Lambda/\Gamma) \to H_1(\Lambda/\Gamma, \mathbb{Z}_2)$. Let

$$A = J^{-1}(a') \cap \Gamma^*$$

and let $\mathrm{Tr}(A)$ denote the trace of a matrix $A \in \mathrm{SL}(2, \mathbb{R})$. Set

$$\omega_{\Gamma^*}(a) = \begin{cases} 0 & \text{if } \mathrm{Tr}(A) < 0, \\ 1 & \text{if } \mathrm{Tr}(A) > 0. \end{cases}$$

By Theorem 7.2 of Chapter 1, the mapping $\Gamma^* \mapsto \omega_{\Gamma^*}$ establishes a one-to-one correspondence between liftings of the group $\Gamma$ and Arf functions on $P = \Lambda/\Gamma$.

**2.** Now consider the group

$$\mathrm{SL}_\pm(2, \mathbb{R}) = \{A \in GL(2, \mathbb{R}) \mid \det A = \pm 1\}.$$

We extend the projection $J$ to a homomorphism $J \colon \mathrm{SL}_\pm(2, \mathbb{R}) \to \widetilde{\mathrm{Aut}}(\Lambda)$ by setting

$$J(A) = \frac{a\bar{z} + b}{c\bar{z} + d} \quad \text{if} \quad A = \begin{pmatrix} a & b \\ c & d \end{pmatrix} \quad \text{and} \quad \det A = -1.$$

Let $\tilde{\Gamma}$ be a given Fuchsian group. We call a subgroup $\tilde{\Gamma}^* \subset \mathrm{SL}_\pm(2, \mathbb{R})$ a *lifting* of $\tilde{\Gamma}$ if $J(\tilde{\Gamma}^*) = \tilde{\Gamma}$ and $J|_{\tilde{\Gamma}^*} \colon \tilde{\Gamma}^* \to \tilde{\Gamma}$ is an isomorphism. It is clear that a lifting $\tilde{\Gamma}^*$ of $\tilde{\Gamma}$ induces a lifting $\Gamma^* = \tilde{\Gamma}^* \cap \mathrm{SL}(2, \mathbb{R})$ of $\Gamma = \tilde{\Gamma} \cap \mathrm{Aut}(\Lambda)$, whence an Arf function $\omega_{\tilde{\Gamma}^*} = \omega_{\Gamma^*} \colon H_1(\Lambda/\Gamma, \mathbb{Z}_2) \to \mathbb{Z}_2$.

LEMMA 4.1. *The Arf function $\omega_{\tilde{\Gamma}^*}$ is a nonspecial Arf function on the real curve $[\tilde{\Gamma}]$.*

Proof. The Arf function $\omega_{\tilde{\Gamma}^*}$ is real, since for any $\alpha \in \tilde{\Gamma}^* \setminus \Gamma^*$, $a' \in \Gamma$ and $a = P_r(a')$ the identity

$$\omega_{\tilde{\Gamma}^*}(\tau a) = \mathrm{Tr}(\alpha(J^{-1}(a') \cap \Gamma^*)\alpha^{-1}) = \mathrm{Tr}(J^{-1}(a') \cap \Gamma^*) = \omega_{\tilde{\Gamma}^*}(a)$$

holds. Let us prove that $\omega_{\tilde{\Gamma}^*}$ is nonspecial. Let $c \subset P \setminus P^\tau$ be a simple contour such that $\tau c = c$, and let $C \subset \Gamma$ be its image under the natural isomorphism $\pi_1(\Lambda/\Gamma, p) \to \Gamma$. Let

$$\tilde{C}^* = J^{-1}(\tilde{C}) \cap \tilde{\Gamma}^* = \begin{pmatrix} a & b \\ c & d \end{pmatrix},$$

where $\tilde{C} = \overline{C}\sqrt{C}$ (see Section 2.4). Then

$$J^{-1}(C) \cap \Gamma^* = (\tilde{C}^*)^2 = \begin{pmatrix} a & b_2 \\ c & d \end{pmatrix}.$$

Therefore,

$$\mathrm{Tr}(J^{-1}(C) \cap \Gamma^*) > 0 \quad \text{and} \quad \omega(c) = 1.$$

$\square$

Two liftings $\tilde{\Gamma}_1^*$ and $\tilde{\Gamma}_2^*$ of a real Fuchsian group $\tilde{\Gamma}$ are said to be *similar* if $(\tilde{\Gamma}_1^* \setminus \Gamma^*) = -(\tilde{\Gamma}_2^* \setminus \Gamma^*)$.

LEMMA 4.2. *Let $\omega$ be a nonspecial Arf function on $[\tilde{\Gamma}]$. Then there are exactly two liftings $\tilde{\Gamma}^*$ of $\tilde{\Gamma}$ such that $\omega_{\tilde{\Gamma}^*} = \omega$, and these two liftings are similar.*

Proof. According to Section 7 of Chapter 1, there is a unique lifting $\Gamma^* \subset \mathrm{SL}(2, R)$ of $\Gamma = \tilde{\Gamma} \cap \mathrm{Aut}(\Lambda)$ such that $\omega_{\Gamma^*} = \omega$. Therefore, each lifting $\tilde{\Gamma}^*$ of $\tilde{\Gamma}$ satisfying the condition $\omega_{\tilde{\Gamma}^*} = \omega$ is generated by $\Gamma^*$ and a matrix $\alpha$ such that $J(\alpha) \in \tilde{\Gamma} \setminus \Gamma$. If $(J(\alpha))z = \frac{a\bar{z}+b}{c\bar{z}+d}$, then $\alpha = \pm \left( \begin{smallmatrix} a & b \\ c & d \end{smallmatrix} \right)$. Since the Arf function $\omega$ is real, we have $\mathrm{Tr}(\alpha A \alpha^{-1}) = \mathrm{Tr}(A)$ for $A \in \Gamma^*$, whence $\alpha\Gamma^*\alpha^{-1} = \Gamma^*$. Therefore, the group $\tilde{\Gamma}^*$ generated by $\Gamma^*$ and $\alpha$ is a lifting of $\tilde{\Gamma}$. $\square$

Lemmas 4.1 and 4.2 imply

THEOREM 4.1 ([**80**]). *The mapping $\tilde{\Gamma}^* \mapsto \omega_{\tilde{\Gamma}^*}$ establishes a one-to-one correspondence between similarity classes of liftings of a real Fuchsian group $\tilde{\Gamma}$ and nonspecial Arf functions on the real curve $[\tilde{\Gamma}]$.*

**3.** The natural isomorphism $\pi_1(\Lambda/\Gamma, p) \to \Gamma$ takes the free homotopy class of a contour $c \in P = \Lambda/\Gamma$ to the conjugacy class $\Gamma_c \subset \Gamma$ independent of the choice of the point $p$. In this way we associate to each simple geodesic contour $c \in P$ a subset $\Gamma_c \subset \Gamma$, and $\Phi(\ell(C)) = c$ provided $C \in \Gamma_c$ and $\Phi \colon \Lambda \to P$ is the natural projection.

Now let $\tilde{\Gamma}$ be a real Fuchsian group and let $c$ be an oval of the curve $(P, \tau) = [\tilde{\Gamma}]$. Take $C \in \Gamma_c$. Replacing the group $\tilde{\Gamma}$ by a conjugate one, if necessary, we may assume that $\ell(C) = I = \{z \in \Lambda \mid \mathrm{Re}\, z = 0\}$. Then $\tilde{\Gamma}$ contains the involution $\beta(z) = -\bar{z}$. The lifting $\tilde{\Gamma} \to \tilde{\Gamma}^*$ takes $\beta$ to a matrix of the form $\sigma \left( \begin{smallmatrix} -1 & 0 \\ 0 & 1 \end{smallmatrix} \right)$, where $\sigma = \pm 1$. Let us equip the half-line $I$ with the orientation of increasing $\mathrm{Im}\, z$ if $\sigma = 1$, and with the opposite orientation if $\sigma = -1$. The projection $\Phi$ pushes this orientation to the contour $c = \Phi(I)$. It is completely determined by the lifting $\tilde{\Gamma}^*$, and we call it the *orientation of an oval induced by the lifting $\tilde{\Gamma}^*$*.

LEMMA 4.3 ([**67**]). *Let $\tilde{\Gamma}^*$ be a lifting of a real Fuchsian group $\tilde{\Gamma}$, $(P, \tau) = [\tilde{\Gamma}]$, let $c_1, c_2$ be ovals of the involution $\tau$ endowed with the orientation induced by $\tilde{\Gamma}^*$, and let $a \subset P$ be a simple oriented contour intersecting $c_1$ and $c_2$ and such that $\tau a = -a$. Then $a$ has coinciding numbers of intersection with $c_1$ and $c_2$ if and only if $\omega_{\tilde{\Gamma}^*}(a) = 1$.*

Proof. Replacing the group $\tilde{\Gamma}$ by a conjugate one, if necessary, we may assume that $\Gamma_a \supset A$, where $A(z) = \lambda z$ and $\lambda > 1$ (see Fig. 2.4.1).

In this case (because of $\tau a = -a$, $c_i \cap a \neq \varnothing$ and $c_i \subset P^\tau$) $\Gamma_{c_i} \supset C_i$, where

$$C_i = \frac{\alpha_i(\lambda_i + 1)z + \alpha_i^2(\lambda_i - 1)}{(\lambda_i - 1)z + \alpha_i(\lambda_i + 1)}, \quad \lambda_i > 1, \quad \overline{C}_i = \frac{\alpha_i^2}{\overline{z}}$$

and $A = \overline{C}_1\overline{C}_2$. Let us set $A^* = J^{-1}(A) \cap \tilde{\Gamma}^*$, $C_i^* = J^{-1}(C_i) \cap \tilde{\Gamma}^*$, $\overline{C}_i^* = J^{-1}(\overline{C}_i) \cap \tilde{\Gamma}^*$. Then by the definition of the orientation induced by $\tilde{\Gamma}^*$, we have $\overline{C}_i^* = -\begin{pmatrix} 0 & \alpha_i \\ \alpha_i^{-1} & 0 \end{pmatrix}$ and, therefore,

$$A^* = \overline{C}_1^*\overline{C}_2^* = -\begin{pmatrix} \alpha_1\alpha_2^{-1} & 0 \\ 0 & \alpha_1^{-1}\alpha_2 \end{pmatrix}.$$

On the other hand, the intersection numbers of $a$ with $c_1$ and $c_2$ coincide if and only if the fixed attracting points $\alpha_1$ and $\alpha_2$ have the same sign. This requirement is equivalent to $\operatorname{Tr}(A^*) > 0$, i.e., $\omega_{\tilde{\Gamma}^*}(a) = 1$.  □

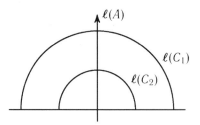

FIGURE 2.4.1

**4.** Now let $c \subset P$ be an invariant contour in $(P, \tau) = [\tilde{\Gamma}]$ such that $c \cap P^\tau = \varnothing$. Take $C \in \Gamma_c$. Replacing $\tilde{\Gamma}$ by a conjugate group, as above, we may assume that $\ell(C) = I$. Therefore, $\tilde{\Gamma}$ contains a mapping of the form $\beta(z) = -\lambda\overline{z}$, where $\lambda > 0$. The lifting $\tilde{\Gamma} \to \tilde{\Gamma}^*$ takes $\beta$ to a matrix of the form

$$\sigma\begin{pmatrix} -\lambda^{1/2} & 0 \\ 0 & \lambda^{-1/2} \end{pmatrix},$$

where $\sigma = \pm 1$. As before, endow $I$ with the increasing $\operatorname{Im} z$ orientation if $\sigma = 1$, and with the opposite orientation if $\sigma = -1$. The projection $\Phi$ pushes this orientation to $c = \Phi(I)$. The resulting orientation depends only on the lifting $\tilde{\Gamma}^*$, and it will be called the *orientation of the invariant contour induced by* $\tilde{\Gamma}^*$.

THEOREM 4.2. *Let* $\tilde{\Gamma}^*$ *be a lifting of a Fuchsian group* $\tilde{\Gamma}$ *and let* $(P, \tau) = [\tilde{\Gamma}]$ *be a real algebraic curve of type* $(g, k, 0)$. *Let* $(c_1, \dots, c_g)$ *be a set of pairwise disjoint simple contours such that* $P^\tau = \bigcup_{i=1}^k c_i$ *and* $\tau(c_i) = c_i$. *Then there is an invariant contour* $c_{g+1}$ *intersecting none of* $c_1, \dots, c_g$ *and cutting together with them the surface* $P$ *into two spheres with holes* $P_1, P_2$ *in such a way that the orientation of the contours* $c_1, \dots, c_g$ *induced by* $\tilde{\Gamma}^*$ *coincides with their orientation as the boundary of one of the surfaces* $P_i$.

Proof. Consider a set of pairwise disjoint invariant contours $c_1, \ldots, c_g, c \subset P$ such that $P^\tau = \bigcup_{i=1}^{k} c_i$ and $P \setminus \left( \bigcup_{i=1}^{g} c_i \cup c \right)$ splits into two spheres with

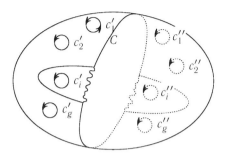

FIGURE 2.4.2

holes. Such a set exists by Lemma 1.2. Equip the contours $c_1, \ldots, c_g$ with the orientation induced by the lifting $\tilde{\Gamma}^*$. Their images in $\tilde{P} = P \setminus \bigcup_{i=1}^{g} c_i$ are pairs of contours of opposite orientation $c_i', c_i''$, where $c_i'$ and $c_j'$ belong to the same connected component of $\tilde{P} \setminus c$ (see Fig. 2.4.2). By a symmetric modification of the contour $c$ as in Fig. 2.4.2, we can transform it into a symmetric contour $c_{g+1}$ separating contours of opposite orientation. $\qquad\square$

## 5. Spinors of rank 1 on real algebraic curves

**1.** Recall that a line bundle $e\colon E \to P$ is called a *spinor bundle of rank* 1 if its tensor square is the cotangent bundle. In the sequel, if not specified otherwise, the words "spinor bundle" mean a spinor bundle of rank 1 over a Riemann surface $P$.

In Section 1 of Chapter 1 we established a one-to-one correspondence between the liftings $\Gamma^*$ of a Fuchsian group $\Gamma$ and the spinor bundles on $P = \Lambda/\Gamma$. The spinor bundle $e_{\Gamma^*}$ corresponding to $\Gamma^*$ has the form

$$(\Lambda \times \mathbb{C})/\Gamma^* \to \Lambda/\Gamma,$$

where $\Gamma^*$ acts on $(\Lambda \times \mathbb{C})$ according to the rule

$$\begin{pmatrix} a & b \\ c & d \end{pmatrix}(z, x) = \left( \frac{az + b}{cz + d}, (cz + d)x \right).$$

Hence, the composite mapping $e_{\Gamma^*} \to \Gamma^* \to \omega_{\Gamma^*}$ establishes a one-to-one correspondence $e \to \omega_e$ between the spinor bundles and the Arf functions on $P = \Lambda/\Gamma$.

**2.** A *spinor bundle on a real curve* $(P, \tau)$ is a pair $(e, \beta)$, where $e\colon E \to P$ is a spinor bundle and $\beta\colon E \to E$ is an antiholomorphic involution such that $e\beta = \tau e$. Two spinor bundles $(e_1, \beta_1)$ and $(e_2, \beta_2)$ on curves $(P_1, \tau_1)$ and $(P_2, \tau_2)$, respectively, are isomorphic if there exist biholomorphic mappings $\phi_E\colon E_1 \to E_2$ and $\phi_P\colon P_1 \to P_2$ such that

$$e_2\phi_E = \phi_P e_1, \quad \beta_2\phi_E = \phi_E\beta_1, \quad \tau_2\phi_P = \phi_P\tau_1.$$

Usually we make no difference between isomorphic bundles.

Now let us associate to each lifting $\tilde{\Gamma}^*$ of a real Fuchsian group $\tilde{\Gamma}$ a spinor bundle $e_{\tilde{\Gamma}^*}$ on the real curve $(P, \tau) = [\tilde{\Gamma}]$. By definition, the bundle $e_{\tilde{\Gamma}^*}$ is $(e_{\Gamma^*}, \beta_{\tilde{\Gamma}^*})$, where $\beta_{\tilde{\Gamma}^*} \colon (\Lambda \times \mathbb{C})/\Gamma^* \to (\Lambda \times \mathbb{C})/\Gamma^*$ is induced by the mapping

$$(z, x) \mapsto \left( \frac{a\bar{z} + b}{c\bar{z} + d}, (c\bar{z} + d)\bar{x} \right), \quad \begin{pmatrix} a & b \\ c & d \end{pmatrix} \in \tilde{\Gamma}^* \setminus \tilde{\Gamma}.$$

LEMMA 5.1. *The mapping $\tilde{\Gamma}^* \mapsto e_{\tilde{\Gamma}^*}$ establishes a one-to-one correspondence between the similarity classes of liftings $\tilde{\Gamma}^*$ of a real Fuchsian group $\tilde{\Gamma}$ and spinor bundles over $(P, \tau) = [\tilde{\Gamma}]$.*

Proof. Let $(e, \beta)$ be an arbitrary spinor bundle over $(P, \tau)$. According to Section 10 of Chapter 1, there is a unique lifting $\Gamma^*$ of the group $\Gamma = \tilde{\Gamma} \cap \mathrm{Aut}(\Lambda)$ such that

$$e \colon (\Lambda \times \mathbb{C})/\Gamma^* \to \Lambda/\Gamma.$$

After replacing, if necessary, the group $\tilde{\Gamma}$ by a conjugate one, we may assume that $\tilde{\Gamma}$ contains a mapping of the form

$$z \mapsto -\mu\bar{z},$$

where $\mu \geqslant 1$. Let $\mu_*$ be the minimal one among all such $\mu$. Let $\nu = \sqrt{\mu_*}$. Then the group $\Gamma^*$ and the matrices $\pm \left( \begin{smallmatrix} -\nu^{-1} & 0 \\ 0 & \nu \end{smallmatrix} \right)$ generate the liftings $\tilde{\Gamma}^*_+$ and $\tilde{\Gamma}^*_-$ of $\tilde{\Gamma}$. These are the only liftings of $\tilde{\Gamma}$ containing $\Gamma^*$. Moreover, $e_{\tilde{\Gamma}^*_\pm} = e$, and the isomorphism between $e_{\tilde{\Gamma}^*_+}$ and $e_{\tilde{\Gamma}^*_-}$ is generated by the involution $(z, x) \mapsto (z, -x)$. $\qquad \square$

Using Lemma 2.1 and Theorem 4.1, we immediately obtain

THEOREM 5.1 ([**67**], [**80**]). *The mapping $e \mapsto \omega_e$ establishes a one-to-one correspondence between spinor bundles and nonspecial Arf functions on $(P, \tau)$.*

Let $(e, \beta)$ be a spinor bundle on a real curve $(P, \tau)$. Using Lemmas 2.1 and 5.1, we construct an isomorphism

$$(e, \beta) \to (e_{\tilde{\Gamma}^*}, \beta_{\tilde{\Gamma}^*}),$$

where $\tilde{\Gamma}^*$ is a lifting of the real Fuchsian group $\tilde{\Gamma}$ and $(P, \tau) = [\tilde{\Gamma}]$. Equip the ovals and the invariant contours of $[\tilde{\Gamma}]$ not intersecting the ovals with the orientation induced by $\tilde{\Gamma}^*$ (see Section 4). Then each spinor bundle $(e, \beta)$ on the real curve $(P, \tau)$ induces an orientation on the ovals and invariant contours of $(P, \tau)$ not intersecting the ovals. This orientation is well defined up to the common inversion of the orientation on all ovals and all invariant contours.

**3.**  A holomorphic section $\eta \colon P \to E$ of a spinor bundle $e \colon E \to P$ is called a *spinor*. A section $\eta$ of an arbitrary spinor bundle $(e, \beta)$ on a real curve $(P, \tau)$ is called a *real spinor* if $\beta\eta = \eta\tau$. Let $\{\tilde{\Gamma}^*_1, \tilde{\Gamma}^*_2\}$ be the similarity class which corresponds, according to Lemma 5.1, to the bundle $(e, \beta)$. Then we may treat $\eta$ as a section of the spinor bundle $e_{\Gamma^*}$, where $\Gamma^* = \tilde{\Gamma}_1^* \cap \tilde{\Gamma}_2^*$. The spinor $\eta$ is invariant with respect to one of the involutions $\beta_{\tilde{\Gamma}_i^*}$ and antiinvariant with respect to the other one. Suppose, for definiteness, that $\beta_{\tilde{\Gamma}_1^*}\eta = \eta\tau$. We call the orientation induced by the lifting $\tilde{\Gamma}_1^*$ on the ovals and invariant contours of $(P, \tau)$ the *orientation induced by* $\eta$.

A local chart $u\colon U \to \mathbb{C}$ in a neighborhood of a real point $p_0 \in P^\tau$ is said to be *real* provided $\tau U = U$ and $u(\tau p) = \overline{u(p)}$. For a real chart, $u(U \cap P^\tau) \subset \mathbb{R}$. We say that a real chart is *consistent with a spinor* $\eta$ if $u$ takes the orientation of the oval $a \ni p_0$ induced by the spinor to the standard orientation of $\mathbb{R} \subset \mathbb{C}$.

A local chart on a Riemann surface determines a local trivialization of the cotangent bundle, whence a local trivialization of the spinor bundle. Hence a spinor is described in a local chart by a function $f(u) \in \mathbb{C}$.

LEMMA 5.2. *Let $(e, \beta)$ be a spinor bundle over a real curve $(P, \tau)$ and let $\eta$ be a real spinor of this bundle. Then in any real chart $u\colon U \to \mathbb{C}$ consistent with $\eta$ the spinor $\eta$ is described by a function $f(u)$ such that $f(u\tau) = \overline{f(u)}$.*

Proof. Set $i_e\colon (z, x) \mapsto (iz, x)$. Lemmas 2.1 and 5.1 allow us to assume that

$$(P, \tau) = [\tilde{\Gamma}], \quad e = (\Lambda \times C)/\Gamma^* \to [\Gamma], \quad \begin{pmatrix} -1 & 0 \\ 0 & 1 \end{pmatrix} \in \tilde{\Gamma}^* \setminus \Gamma^*$$

and $ei_e\colon (-i\Lambda \times 0) \to P$ induces a real chart $u$ consistent with $\eta$. In this chart, the relation $\beta\eta = \eta\tau$ takes the form

$$(u\tau, \overline{f(u)}) = \beta_{\tilde{\Gamma}^*}(u, f(u)) = (u\tau, f(u\tau)),$$

whence $\overline{f(u)} = f(u\tau)$. A passage to any other real chart consistent with $\eta$ preserves this equation. $\qquad\square$

THEOREM 5.2 ([**78**], [**80**]). *Let $(e, \beta)$ be a spinor bundle over a real curve $(P, \tau)$, let $\eta$ be a real spinor of this bundle, and let $a$ be an oval of $(P, \tau)$. Then the parity of the number of zeros of $\eta$ on $a$ is opposite to the parity of $\omega_e(a)$.*

Proof. Lemmas 2.1 and 5.1 allow us to assume that

$$(P, \tau) = [\tilde{\Gamma}], \quad e = (\Lambda \times \mathbb{C})/\Gamma^* \to [\Gamma],$$

$$\begin{pmatrix} -1 & 0 \\ 0 & 1 \end{pmatrix} \in \tilde{\Gamma}^* \setminus \Gamma^*, \quad a = I/\Gamma, \quad \text{where } I = \{z \in \Lambda \mid \operatorname{Re} z = 0\}.$$

In a local chart $u$ induced by the projection $e\colon (\Lambda \times 0) \to P$ the spinor $\eta$ has the form $(u, f(u))$, where $u \in \Lambda$ and $f(u)$ is a holomorphic function such that

$$f\left(\frac{\alpha u + \beta}{\gamma u + \delta}\right) = f(u)(\gamma u + \delta)$$

for an arbitrary matrix $\begin{pmatrix} \alpha & \beta \\ \gamma & \delta \end{pmatrix} \in \Gamma^*$.

Corresponding to the contour $a$ is the matrix

$$A = \sigma(a) \begin{pmatrix} \lambda & 0 \\ 0 & \lambda^{-1} \end{pmatrix} \in \Gamma^*,$$

where

$$\sigma(a) = \begin{cases} 1 & \text{if } \omega(a) = 1, \\ -1 & \text{if } \omega(a) = 0. \end{cases}$$

Thus $f(\lambda^2 u) = \sigma(a)f(u)$. Moreover, the natural projection $\Lambda \to \Lambda/\Gamma$ establishes a one-to-one correspondence between the interval $(v, \lambda^2 v] \in I$ and the contour $a$. Therefore, the number of zeros of $\eta$ on $a$ coincides with the number of zeros of $f(u)$ on the interval $(v, \lambda^2 v] \in I$. On the other hand, the mapping $e\colon (\Lambda \times 0) \to P$ induces real charts in a neighborhood of each point of $a$, hence, by Lemma 5.2, the function $f(u)$ is real in $(v, \lambda^2 v] \in I$. Therefore, the number of zeros of $f$ on the interval $(v, \lambda^2 v] \in I$ is even provided $\sigma(a) = 1$, and odd if $\sigma(a) = -1$. $\qquad\square$

**4.** The following theorem holds.

THEOREM 5.3 ([**78**], [**80**]). *Let $c_1, \ldots, c_k$ be the oriented ovals of a real algebraic curve $(P, \tau)$ of type $(g, k, 0)$. Let $0 \leqslant m \leqslant k$, $\alpha_1, \ldots, \alpha_k \in \mathbb{Z}_2$, and $\sum_{i=1}^{k} \alpha_i \equiv g + 1 \pmod 2$. Then there is a real spinor $\eta$ on $(P, \tau)$ such that (1) the orientation of $c_i$ induced by $\eta$ coincides with the initial orientation if and only if $i \leqslant m$; (2) the parity of the number of zeros of $\eta$ on $c_i$ is $\alpha_i$.*

Proof. By Theorem 1.2, there is a set of pairwise disjoint contours $c_1, \ldots, c_{g+1}$ that are invariant with respect to $\tau$ and cut $P$ into two surfaces with holes $P_1$ and $P_2$. The orientation of $P_1$ induces some new orientation on $\varphi P_1 = \{c_1, \ldots, c_{g+1}\}$. Without loss of generality we may assume that it coincides with the original one on $c_1$. Let us connect the contours $c_{g+1}$ and $c_i$ by a segment $\ell_i \subset P_1$ and consider the simple contour $d_i = \ell_i \cup \tau \ell_i \cup r_i \cup r_{g+1}$, where $r_j \subset c_j$. Set $\omega(c_i) = 1 - \alpha_i$ for $i \leqslant k$ and $\omega(c_i) = 1$ for $k < i \leqslant g$. For $1 \leqslant i \leqslant m$, we set $\omega(d_i) = 0$ if and only if the orientation induced by $P_1$ is opposite to the initial orientation of $c_i$. For $m < i \leqslant k$, we set $\omega(d_i) = 0$ if and only if the orientation induced by $P_1$ is opposite to the initial orientation of $c_i$. For $k < i \leqslant g$ we set $\omega(d_i) = 0$. The function $\omega$ admits a unique extension as an Arf function $\omega\colon H_1(P, \mathbb{Z}_2) \to \mathbb{Z}_2$ and $\omega(c_{g+1}) = 1$ because of $\sum_{i=1}^{k} \alpha_i \equiv g + 1 \pmod 2$. Moreover, $\tau d_i = -d_i + c_{g+1} + \tilde{c}_i$, where

$$\tilde{c}_i = \begin{cases} 0 & \text{for } i \leqslant k, \\ c_i & \text{for } i > k. \end{cases}$$

Hence $\omega(\tau d_i) = \omega(-d_i + c_{g+1} + \tilde{c}_i) = \omega(d_i)$, and therefore, $\omega$ is a real Arf function. By Lemma 3.3, this Arf function is nonspecial. By Lemma 2.1, $(P, \tau) = [\tilde{\Gamma}]$, where $\tilde{\Gamma}$ is a real Fuchsian group. By Lemma 4.2, $\omega = \omega_{\tilde{\Gamma}^*}$, where $\tilde{\Gamma}^*$ is some lifting of $\tilde{\Gamma}$. By definition, $\omega_e = \omega$, where $(e, \beta) = (e_{\Gamma^*}, \beta_{\tilde{\Gamma}^*})$.

Consider together with $\omega$ the real Arf function $\omega'$ such that $\omega'(c_i) = \omega(c_i)$, $\omega'(d_i) = 1 - \omega(d_i)$. Corresponding to this function is the real spinor bundle $(e', \beta')$ such that $\omega_{e'} = \omega'$. Moreover,

$$\delta(\omega) + \delta(\omega') = \sum_{i=1}^{g} \omega(c_i) = 1,$$

since $\sum_{i=1}^{k} \alpha_i \equiv g + 1 \pmod 2$. Therefore, either $\delta(\omega) = 1$ or $\delta(\omega') = 1$. Suppose for definiteness that $\delta(\omega_e) = \delta(\omega) = 1$. According to [4] and [51], this implies that $e$ has a holomorphic section $\xi$. Then one of the sections $\eta = \xi + \beta\xi$ or $\eta = i(\xi - \beta\xi)$ is a real nonzero section of $(e, \beta)$. By Lemma 4.3 and Theorem 5.2, this section satisfies the assertion of Theorem 5.3.          $\square$

THEOREM 5.4 ([78], [80]). *Let $(P, \tau)$ be a real algebraic curve of type $(g, k, 1)$. Orient the ovals $c_1, \ldots, c_k$ of this curve as the boundary of a connected component $P_1$ of the set $P \setminus P^\tau$. Consider a set $\alpha_1, \ldots, \alpha_k \in \mathbb{Z}_2$ containing evenly many zeros and such that $\alpha_1 = \alpha_k = 0$. Let $1 \leqslant m < k$ and $\sum_{i=1}^{m} \alpha_i \equiv m + 1 \pmod 2$. Then there is a real spinor $\eta$ on $(P, \tau)$ such that (1) the orientation of $c_i$ induced by $\eta$ coincides with the original one if and only if $i \leqslant m$; (2) the parity of the number of zeros of $\eta$ on $c_i$ is $\alpha_i$.*

Proof. Let us connect the ovals $c_i$ and $c_k$ by a segment $\ell_i \subset P_1$ and set $d_i = \ell_i \cup \tau\ell_i$ $(i = 1, \ldots, k-1)$. Consider an arbitrary Arf function $\omega_1$ on $P_1$ such that $\omega_1(c_i) = 1 - \alpha_i$ (such a function exists according to Lemma 8.1 of Chapter 1). Let us extend $\omega_1$ as an Arf function $\omega$ on $P$ assuming that $\omega(\tau w) = \omega(w)$ for $w \in H_1(P_1, \mathbb{Z}_2)$ and $\omega(d_i) = 1$ if and only if $i \leqslant m$. Then $\delta(\omega) = 1$. The rest of the proof coincides with the corresponding part of the proof of Theorem 5.3.          $\square$

## 6. Holomorphic differentials on real algebraic curves

**1.**          In this section we assume that all ovals of a real algebraic curve $X = (P, \tau)$ are endowed with an orientation. This orientation is induced by the orientation of one of the connected components of the surface $P \setminus P^\tau$ provided $\varepsilon(X) = 1$. A real chart $u \colon U \to \mathbb{C}$ is *consistent* with the orientation of $P^\tau$ if $u$ takes the oriented segment $\ell = U \cap P^\tau$ to a segment $u(\ell) \subset \mathbb{R}$ endowed with the standard orientation.

Recall that a *holomorphic differential* on a Riemann surface $P$ is a section $\xi \colon P \to T^*$ of the cotangent bundle $t \colon T^* \to P$. Now let $(P, \tau)$ be a real algebraic curve. The involution $\tau$ induces an antiholomorphic involution $\tau^* \colon T^* \to T^*$ such that $t\tau^* = \tau t$. A differential $\xi$ is said to be *real* if $\tau^*\xi = \xi\tau$. In a real chart, it has the form $\xi = f(u)du$, where $f(\overline{u}) = \overline{f(u)}$. In particular, $f(u(p)) \in \mathbb{R}$ for $p \in P^\tau$. The sign of $f(u(p)) \in \mathbb{R}$ is independent of the choice of the real chart consistent with the orientation of $P^\tau$, and we call this sign the sign of $\xi$ at $p \in P^\tau$.

We say that a real differential $\xi$ is *positive* (respectively, *nonnegative, nonpositive, negative*) on an oval $a \in P^\tau$ if it is positive (respectively, nonnegative, nonpositive, negative) at each point of the oval.

LEMMA 6.1. *Let $\eta$ be a real spinor on the curve $(P, \tau)$. Then $\xi = \eta^2$ is a real differential, and it is nonnegative on an oval $a \in P^\tau$ if the orientation of $a$ induced by $\eta$ coincides with the initial one, and it is nonpositive if the orientation induced by $\eta$ is opposite to the initial one.*

Proof. If the spinor $\eta$ is described in a real chart $u \colon U \to \mathbb{C}$ consistent with the orientation of $P^\tau$ by a function $f(u)$, then $\xi = f^2(u) du$. If the orientation of $a$ is induced by $\eta$, then, by Lemma 5.2, $f(u\tau) = \overline{f(u)}$ and $f^2$ is nonnegative on $a$. The inversion of the orientation of the oval changes the sign of $f^2$.                                                                    □

THEOREM 6.1 ([**65**], [**80**]). *Let $(P, \tau)$ be a real algebraic curve of type $(g, k, \varepsilon)$ with the ovals $c_1, \ldots, c_k$, where $k = k_+ + k_- + k_0$, $k_0 < g$, and $k_+ \cdot k_- \neq 0$ for $\varepsilon = 1$. Then there is a real differential on $(P, \tau)$ which is nonnegative on $c_i$ for $i \leqslant k_+$, nonpositive on $c_i$ for $k_+ < i \leqslant k_+ + k_-$, and has zeros on $c_i$ for $i > k_+ + k_-$.*

Proof. By Theorems 5.3 and 5.4, there is a real spinor $\eta$ having zeros on $c_{k_+ + k_- + 1}, \ldots, c_k$ and inducing on the other ovals $c_i$ the orientation coinciding with the orientation of $P^\tau$ for $i \leqslant k_+$ and opposite to that of $P^\tau$ for $k_+ < i \leqslant k_+ + k_-$. Then, by Lemma 6.1, the differential $\xi = \eta^2$ possesses all the desired properties.                                                                    □

**2.**    Consider real *M-curves*, that is, curves of type $(g, g+1, 1)$, in more detail.

LEMMA 6.2. *Let $c_1, \ldots, c_{g+1}$ be the ovals of an $M$-curve of genus $g$ and let $1 \leqslant \alpha \leqslant n < \beta \leqslant g+1$. Then there is a real differential $\xi_1$ which is positive on $c_\alpha$, nonnegative on $c_1, \ldots, c_n$, negative on $c_\beta$ and nonpositive on $c_{n+1}, \ldots, c_{g+1}$.*

Proof. By Theorem 5.4, there is a real spinor $\eta$ which induces the orientation coinciding with the initial one on $c_1, \ldots, c_n$, the orientation opposite to the initial one on $c_{n+1}, \ldots, c_{g+1}$, and has zeros on the ovals $c_i$, where $i \neq \alpha, \beta$. But the total number of zeros of a spinor is $g - 1$ [**51**]. Therefore, $\eta$ has no zeros on the ovals $c_\alpha$ and $c_\beta$. Now it follows from Lemma 6.1 that the real differential $\xi = \eta^2$ satisfies all the assertions of the lemma.           □

Now we immediately obtain

LEMMA 6.3. *Let $c_1, \ldots, c_{g+1}$ be the ovals of an $M$-curve of genus $g$ and let $1 \leqslant n < g+1$. Then there is a real differential $\xi$ positive on $c_1, \ldots, c_n$ and negative on $c_{n+1}, \ldots, c_{g+1}$.*

LEMMA 6.4. *Let $\alpha_1 < \cdots < \alpha_{2g+2}$ be real points, $h(x) = \prod_{i=1}^{2g+2}(x - \alpha_i)$, let $P$ be the Riemann surface of the algebraic curve $y^2 = h(x)$, and let $\tau \colon P \to P$ be the antiholomorphic involution induced by the mapping $(x, y) \mapsto (\bar{x}, \bar{y})$. Then $(P, \tau)$ is a real $M$-curve of genus $g$ such that all real differentials of it are positive on one of the ovals.*

Proof. The ovals of $(P, \tau)$ are in one-to-one correspondence with the segments $[\alpha_{2i-1}, \alpha_{2i}]$. Real differentials on $(P, \tau)$ have the form

$$\xi_f = \frac{f(x)dx}{\sqrt{h(x)}},$$

where $f$ is a polynomial with real coefficients of degree at most $g - 1$. If $f(x) > 0$, then the differential has opposite signs on ovals corresponding to neighboring segments. Therefore, if the differential $\xi_f$ is not strictly positive on one of the ovals, then $f$ has more than $g - 1$ zeros, which is impossible, since $\deg f \leqslant g - 1$. $\square$

THEOREM 6.2 ([65]). *For each real differential on an $M$-curve, there is an oval where it is positive, and an oval where it is negative.*

Proof. Let $\tilde{M}$ be the set of all $M$-curves of genus $g$ with an ordered set of ovals $c_1, \ldots, c_{g+1}$. Consider the vector bundle $\tilde{e}: \tilde{E} \to \tilde{M}$ whose fiber $\tilde{e}^{-1}(P, \tau)$ consists of all real differentials on $(P, \tau)$. Consider the basis of a fiber $\tilde{e}^{-1}(P, \tau)$ consisting of the differentials $\xi_i = \xi_i(P, \tau)$ such that $\oint_{c_i} \xi_j = \delta_{ij}$ $(i, j \leqslant g)$. The mapping $\xi_i(P, \tau) \mapsto \xi_i(P', \tau')$ determines a connection $F$ on $\tilde{e}$.

We call a real differential a *differential of type $A$* (respectively, of type $B$) if each oval contains points where the differential is nonpositive (respectively, negative). Let $M^A$ (respectively, $M^B$) be the set of $M$-curves admitting a differential of type $A$ (respectively, of type $B$). Then $M^A$ is a closed set. It is easy to prove, using the connection $F$, that the set $M^B$ is open. Moreover, $M^A \supset M^B$. Let us prove that $M^A \subset M^B$. Let $(P, \tau) \subset M^A$ and let $\xi$ be a differential of type $A$ on $(P, \tau)$. Because of the identity

$$\sum_{i=1}^{g+1} \int_{c_i} \xi = 0,$$

the differential must be negative at least at one point. Let $c$ be an oval containing such a point. By Lemma 6.3, there is a real differential $\gamma$ positive on $c$ and negative on all other ovals. The differential $\xi + \alpha\gamma$ is of type $B$ for $\alpha$ small enough. Hence, $M^A = M^B$ is an open and closed set in $\tilde{M}$. But, by Theorem 2.1, the set $\tilde{M}$ is connected; therefore, if $M^A \neq \varnothing$, then $M^A = \tilde{M}$. The last equality contradicts Lemma 6.4, which states that the set $\tilde{M} \setminus M^A$ contains hyperelliptic curves. Hence $M^A = \varnothing$, i.e., any real differential on an $M$-curve is positive on one of the ovals. The fact that it is also negative on one of the ovals is proved in the same vein. $\square$

THEOREM 6.3 ([65]). *Let $1 \leqslant k \leqslant g + 1 \pmod{2}$, $k \equiv g + 1 \pmod{2}$ and $m \geqslant k - \left[\frac{k}{2}\right]$. Then there exists a real algebraic curve of type $(g, k, 1)$, with ovals $c_1, \ldots, c_k$, such that each real differential that does not have zeros on the ovals $c_1, \ldots, c_m$ is necessarily positive on one of the ovals of the real curve and negative on another oval.*

Proof. Consider the Riemann surface of the curve

$$y^4 - 2y^2[(x - \beta_1) \cdots (x - \beta_m) - (x - \alpha_1) \cdots (x - \alpha_n)]$$
$$+ [(x - \beta_1) \cdots (x - \beta_m) + (x - \alpha_1) \cdots (x - \alpha_n)]^2 = 0,$$

where $\alpha_1 < \cdots < \alpha_n \leqslant \beta_1 < \cdots < \beta_m \in \mathbb{R}$, $n > 0, n, m \equiv 0 \pmod 2$. It is obtained by resolving the singularities of the set

$$((x, y) \in \overline{\mathbb{C}}^2 \mid y = \pm \sqrt{(x - \alpha_1) \cdots (x - \alpha_n)} \pm \sqrt{-(x - \beta_1) \cdots (x - \beta_m)}).$$

The mappings

$$\tau \colon (x, y) \mapsto (\bar{x}, \bar{y}),$$

$$\tau_\alpha \colon (x, \pm \sqrt{-(x - \alpha_1) \cdots (x - \alpha_n)} \pm \sqrt{(x - \beta_1) \cdots (x - \beta_m)})$$
$$\mapsto (x, \mp \sqrt{(x - \alpha_1) \cdots (x - \alpha_n)} \pm \sqrt{-(x - \beta_1) \cdots (x - \beta_m)})$$

and

$$\tau_\beta \colon (x, \pm \sqrt{-(x - \alpha_1) \cdots (x - \alpha_n)} \pm \sqrt{(x - \beta_1) \cdots (x - \beta_m)})$$
$$\mapsto (x, \pm \sqrt{(x - \alpha_1) \cdots (x - \alpha_n)} \mp \sqrt{-(x - \beta_1) \cdots (x - \beta_m)})$$

are commuting involutions on $P$.

It is easy to see that $(P, \tau)$ is a real algebraic curve of type $(g, k, 1)$, where

$$g = \begin{cases} n + m - 1 & \text{for } \alpha_n < \beta_1, \\ n + m - 2 & \text{for } \alpha_n = \beta_1, \end{cases}$$

and

$$k = \begin{cases} n & \text{for } \alpha_n < \beta_1, \\ n - 1 & \text{for } \alpha_n = \beta_1. \end{cases}$$

The involution $\tau_\alpha$ preserves each oval, while the involution $\tau_\beta$ transposes all ovals in pairs if $\alpha_n < \beta_1$ and preserves a single oval provided $\alpha_n = \beta_1$. Let us number the ovals $c_1, \ldots, c_k$ in such a way that $\tau c_i = c_{k+1-i}$. Suppose there is a real differential $\xi$ that is positive on the ovals $c_1, \ldots, c_{n/2}$, and is not negative on the other ovals. Then the differential $\xi + \xi\beta$ is nonnegative on each oval. The involution $\tau$ induces an antiholomorphic involution $\tilde{\tau} \colon \tilde{P} \to \tilde{P}$ on the surface $\tilde{P} = P/\langle \beta \rangle$. It is easy to see that $(\tilde{P}, \tilde{\tau})$ is an $M$-curve of genus $\frac{n}{2} - 1$. The differential $\xi + \xi\beta$ induces a real differential on the curve $(\tilde{P}, \tilde{\tau})$, which is not negative on any of the ovals. The contradiction with Theorem 6.2 thus obtained proves that such a $\xi$ does not exist.  $\square$

## 7. Analogs of Fourier series and the Sturm–Hurwitz theorem on real algebraic curves of arbitrary genus

**1.**   The simplest real algebraic curve is the Riemann sphere $\overline{\mathbb{C}} = \mathbb{C} \cup \infty$ endowed with the antiholomorphic involution $\tau_{\mathbb{C}} \colon z \mapsto \frac{1}{\bar{z}}$. The curve $(\overline{\mathbb{C}}, \tau_{\mathbb{C}})$ has the unique oval

$$c = \{z \in \mathbb{C} \mid |z| = 1\} = \{e^{i\psi} \mid \psi \in \mathbb{R}\}.$$

We shall consider meromorphic functions $f \colon \overline{\mathbb{C}} \to \overline{\mathbb{C}}$ such that $f(\tau_{\mathbb{C}} z) = \overline{f(z)}$. The simplest such functions are the functions holomorphic outside $0$ and $\infty$. They can be represented as Fourier series

$$f(z) = \sum_{n=0}^{\infty} (a_n c_n(z) + b_n s_n(z)),$$

where

$$c_n(z) = \frac{1}{2}(z^n + z^{-n})$$

and

$$s_n(z) = \frac{1}{2i}(z^n - z^{-n}).$$

Restricting $s_n$ and $c_n$ to $c$ we obtain the classical trigonometric functions

$$s_n(e^{i\psi}) = \sin n\psi, \quad c_n(e^{i\psi}) = \cos n\psi.$$

**2.** Now let $(P, \tau)$ be a real algebraic curve of type $(g, k, 1)$ with two points $p_+, p_- \in P \setminus P^\tau$ such that $\tau p_+ = p_-$. Instead of functions, we shall consider tensors of integer and half-integer weight $\lambda$, i.e., sections of the line bundle $\tilde{P} = P \setminus (p_+ \cup p_-)$ meromorphic on $P$ and holomorphic on $E^{\otimes 2\lambda} \to \tilde{P}$, where$(E, \beta)$ is a real spinor bundle on $(\tilde{P}, \tau)$ and $2\lambda \in \mathbb{Z}$. Let $M_\lambda$ denote the space of such tensors. According to [**43**], if $\lambda \neq 0, 1$ or $|n| > g/2$, then for any integer $n + g/2$ there is a unique tensor $f_n^\lambda \in M_\lambda$ with the asymptotics

$$f_n^\lambda = z_\pm^{\pm n - s}(1 + O(z_\pm))(dz_\pm)^\lambda$$

in the neighborhoods $z_\pm$ of the points $p_\pm$, where $s = s(\lambda, g) = g/2 - \lambda(g - 1)$. The involutions $\beta$ and $\tau$ induce the involutions $\beta_\lambda \colon E^{\otimes 2\lambda} \to E^{\otimes 2\lambda}$ and $\tau_\lambda \colon M_\lambda \to M_\lambda$, where $\tau_\lambda f(p) = \beta_\lambda f(\tau p)$. It is easy to see that $\tau_\lambda f_n^\lambda = f_{-n}^\lambda$. A tensor $\xi \in M_\lambda$ is said to be real if $\tau_\lambda \xi = \xi$. In a real local chart, such a tensor has real values on $P^\tau$.

The analogs of the functions $\cos nx$ and $\sin nx$ are the real tensors

$$c_n^\lambda = \frac{1}{2}(f_n^\lambda + f_{-n}^\lambda) \quad \text{and} \quad s_n^\lambda = \frac{1}{2i}(f_n^\lambda - f_{-n}^\lambda),$$

where $n \geqslant 0$. The following theorem is the analog of the summation theorem for trigonometric functions.

THEOREM 7.1. *Let* $\lambda_1, \lambda_2, \lambda_1 + \lambda_2 \neq 0, 1$ *or* $(n_1 + n_2) > g$. *Then*

$$c_{n_1}^{\lambda_1} c_{n_2}^{\lambda_2} - s_{n_1}^{\lambda_1} s_{n_2}^{\lambda_2} = \sum_{n=-g/2}^{g/2} \delta_n c_{n_1 + n_2 - n}^{\lambda_1 + \lambda_2},$$

$$c_{n_1}^{\lambda_1} s_{n_2}^{\lambda_2} - c_{n_2}^{\lambda_2} s_{n_1}^{\lambda_1} = \sum_{n=-g/2}^{g/2} \eta_n s_{n_1 + n_2 - n}^{\lambda_1 + \lambda_2},$$

*where* $\delta_n, \eta_n \in \mathbb{R}$.

Proof. According to [43], we have

$$f_n^\lambda f_m^\mu = \sum_{k=-g/2}^{g/2} Q_{n,m}^{\lambda,\mu,k} f_{n+m-k}^{\lambda+\mu}.$$

The identity $\tau_\lambda f_n^\lambda = f_{-n}^\lambda$ implies

$$Q_{-n,-m}^{\lambda,\mu,k} = \overline{Q}_{n,m}^{\lambda,\mu,-k}.$$

Hence,

$$
\begin{aligned}
c_{n_1}^{\lambda_1} c_{n_2}^{\lambda_2} - s_{n_1}^{\lambda_1} s_{n_2}^{\lambda_2} &= \frac{1}{4}(f_{n_1}^{\lambda_1} + f_{-n_1}^{\lambda_1})(f_{n_2}^{\lambda_2} + f_{-n_2}^{\lambda_2}) \\
&\quad + \frac{1}{4}(f_{n_1}^{\lambda_1} - f_{-n_1}^{\lambda_1})(f_{n_2}^{\lambda_2} - f_{-n_2}^{\lambda_2}) \\
&= \frac{1}{2}(f_{n_1}^{\lambda_1} f_{n_2}^{\lambda_2} + f_{-n_1}^{\lambda_1} f_{-n_2}^{\lambda_2}) \\
&= \frac{1}{2} \sum_{n=-g/2}^{g/2} (Q_{-n_1,-n_2}^{\lambda_1,\lambda_2,n} + \overline{Q}_{n_1,n_2}^{\lambda_1,\lambda_2,n})(f_{n_1+n_2-n}^{\lambda_1+\lambda_2} + f_{-n_1-n_2+n}^{\lambda_1+\lambda_2}) \\
&= \sum_{n=-g/2}^{g/2} \delta_n c_{n_1+n_2-n}^{\lambda_1+\lambda_2},
\end{aligned}
$$

where $\delta_n = 2 \operatorname{Re} Q_{n_1,n_2}^{\lambda_1,\lambda_2,n}$. The second assertion is proved similarly. $\square$

The following theorem is the analog of the Fourier theorem.

THEOREM 7.2 ([79]). *Each real tensor $f^\lambda$ of weight $\lambda \neq 0, 1$ has a unique representation in the form*

$$f^\lambda = \sum_{k=0}^{\infty} (a_k c_k^\lambda + b_k s_k^\lambda),$$

*where $a_k, b_k \in \mathbb{R}$.*

Proof. According to [43], we have

$$f^\lambda = \sum_{n=0}^{\infty} (\alpha_n f_n^\lambda + \beta_n f_{-n}^\lambda) = \sum_{n=0}^{\infty} (a_n c_n^\lambda + b_n s_n^\lambda),$$

where $a_n = \alpha_n + \beta_n$ and $b_n = i(\alpha_n - \beta_n)$. The identity $\tau f^\lambda = f^\lambda$ implies $\beta_n = \bar{\alpha}_n$, whence $a_n, b_n \in \mathbb{R}$. $\square$

The following theorem is the analog of the classical Sturm–Hurwitz theorem [33].

THEOREM 7.3 ([79]). *Let $\lambda \neq 0, 1$ or $n > \frac{g}{2}$. Then the real tensor*

$$F = \sum_{k=n}^{\infty} (a_k c_k^\lambda + b_r s_k^\lambda)$$

*has at least $2n - g$ zeros on the ovals of $P^\tau$.*

Proof. Let $D$ be the divisor of the tensor $c_n^\lambda$. It has the form $D = D_1 + D_0 + D_2$, where $D_0 \in P^\tau$, $P \setminus P^\tau = P_1 \cup P_2$, $D_i \in P_i$, $\tau D_1 = D_2$. Let $p_+ \in P_1$, let $n_0$ denote the degree of $D_0$, and let $n_1$ denote the degree of

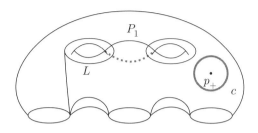

FIGURE 2.7.1

$D_1$. Set $G = \sum_{k-n}^{\infty} \alpha_k f_k^\lambda$, where $\alpha_k = \frac{1}{2}(a_k - ib_k)$. Consider a system of disjoint arcs and contours $L \subset P_1$ such that $Q = P_1 \setminus L$ is simply connected (see Fig. 2.7.1). Let $c \subset Q$ be a simple contour in $Q - p_+$ not homotopic to 0. In the domain bounded by $c$ the function $f = G/c_n^\lambda$ has a zero at $p_+$ of order $2n$, and at most $n_1$ poles. Therefore, the contour $f(c)$ encircles 0 at least $2n - n_1$ times and, therefore, intersects $\operatorname{Im} \mathbb{C} = \{z \in \mathbb{C} \mid \operatorname{Re} z = 0\}$ at least $2(2n - n_1)$ times. We conclude that as $c$ tends to the boundary of $Q$, $f(P^\tau)$ intersects $\operatorname{Im} \mathbb{C}$ at least $4n - (2n_1 + n_0) = 2n - g$ times. But if $p \in P^\tau$ and $f(p) \in \operatorname{Im} \mathbb{C}$, then

$$F(p) = G(p) + (\tau G)(p) = c_1^\lambda(p)(f(p) + \overline{f(\tau(p))}) = 0.$$

Therefore, $F$ has at least $2n - g$ zeros on $P^\tau$. $\qquad \square$

**Remark.** In the case $g = \lambda = 0$ Theorem 7.3 was proved by A. Hurwitz in [**33**]. In this case it got various proofs because of important applications it found in singularity theory. The proof given here is similar to the interpretation of the original Hurwitz proof in the case $g = \lambda = 0$ suggested by V. I. Arnold.

## 8. Jacobians and $\theta$-functions of real algebraic curves

**1.** Let us recall some facts from the classical theory of Riemann surfaces [**26**]. Let $P$ be a compact Riemann surface of genus $g$. A basis in the homology

$$\{a_i, b_i \ (i = 1, \dots, g)\} \in H_1(P, \mathbb{Z})$$

is said to be *symplectic* if the intersection numbers of the cycles are

$$(a_i, a_j) = (b_i, b_j) = 0, \quad (a_i, b_j) = \delta_{ij}.$$

We say that a basis $\xi_1, \dots, \xi_g$ of the space of holomorphic differentials on $P$ is *induced* by a symplectic basis $\{a_i, b_i\}$ if $\oint_{a_k} \xi_j = 2\pi i \delta_{kj}$. In this case the matrix $B = (B_{kj})$, where $B_{kj} = \oint_{b_k} \xi_j$, is symmetric and has a negative definite real part $\mathrm{Re}\, B = (\mathrm{Re}\, B_{ij})$. This property allows one to define the $\theta$-function $\theta\colon \mathbb{C}^g \to \mathbb{C}$ by the identity

$$\theta(z) = \theta(z\,|\,B) = \sum_{N \in \mathbb{Z}^g} \exp\left\{\frac{1}{2}\langle BN, N\rangle + \langle N, z\rangle\right\},$$

where

$$\langle (x_1, \ldots, x_g), (y_1, \ldots, y_g)\rangle = \sum_{i=1}^{g} x_i y_i.$$

Let $G$ be the group spanned by the vectors

$$\ell_k = 2\pi i(\delta_{k1}, \ldots, \delta_{kg}) \quad \text{and} \quad h_k = (B_{k1}, \ldots, B_{kg}) \quad (k = 1, \ldots, g).$$

The complex torus $J = J(P) = \mathbb{C}^g/G$ is called the *Jacobian* of $P$. Let $\Phi\colon \mathbb{C}^g \to J$ be the natural projection.

An unordered set of $k$ points on $P$ is called a (*positive*) *divisor of degree $k$*. Let $S_k$ denote the set of all positive divisors of degree $k$. Choose a point $q$ on $P$ and associate to a divisor $D = \sum_{i=1}^{k} p_i$ the following point of the Jacobian:

$$A_q(D) = \Phi\left(\int_q^D \xi_1, \ldots, \int_q^D \xi_g\right) = \Phi \sum_{i=1}^{k}\left(\int_q^{p_i} \xi_1, \ldots, \int_q^{p_i} \xi_g\right).$$

Then $A_q(S_g) = J$ and the *Abel mapping* $A_q$ is invertible at a general point. The image $K_q$ in $J$ of the vector $(K_q^1, \ldots, K_q^g)$, where

$$K_q^j = \left(\frac{2\pi i + B_{jj}}{2} - \frac{1}{2\pi i} \sum_{\ell \neq j} \int_{a_\ell} (\xi_\ell(p) \int_q^p \xi_j\right),$$

is called the *vector of Riemann constants*. Here $2K_q = -A_q(D_\xi)$, where $D_\xi$ is the divisor of zeros of an arbitrary (holomorphic) differential $\xi$ on $P$. The set

$$(\theta) = A_q(S_{g-1}) + K_q \in J$$

coincides with the image on $J$ of the set of zeros of the $\theta$-function and is called the $\theta$-*divisor*. A subset $\Sigma \subset J$ is said to be *singular* if $\Sigma \cap A_q(S_{g-1}) \neq \varnothing$. For a singular subset, the set $\Sigma + K_q$ contains a zero of the $\theta$-function.

**2.** Now let $(P, \tau)$ be a real algebraic curve. In this and the next section we restrict ourselves to curves with real points. Let $q \in P^\tau$ be such a real point. We shall make use of a symplectic basis

$$\{a_i, b_i \ (i = 1, \ldots, g)\} \subset H_1(P, \mathbb{Z})$$

consistent with $\tau$. Such a basis will be called a *real homology basis*. For a curve of type $(g, k, 0)$, such a basis possesses the following properties: (1) $\tau(a_i) = a_i \ (i = 1, \ldots, g)$, $\tau(b_i) = -b_i \ (i = 1, \ldots, k-1)$, $\tau(b_i) = -b_i + a_i$ $(i = k, \ldots, g)$; (2) the oval containing the point $q$ is homologous to $\sum_{i=1}^{g} a_i$. For a curve of type $(g, k, 1)$, such a basis possesses the following properties: (1) $\tau(a_i) = a_i$, $\tau(b_i) = -b_i \ (i = 1, \ldots, k-1)$, $\tau(a_i) = a_{i+m}$, $\tau(b_i) = -b_{i+m}$

$(i = k, \ldots, k + m - 1)$, where $m = \frac{1}{2}(g + 1 - k)$; (2) the oval containing the point $q$ is homologous to $\sum_{i=1}^{k-1} a_i$.

LEMMA 8.1. *A real basis exists.*

Proof. Let $(P, \tau)$ be a real curve of type $(g, k, 0)$. Then, by Lemma 1.2, there is a set of pairwise disjoint contours $a_0, a_1, \ldots, a_g$ such that

$$\tau(a_i) = a_i, \quad P^\tau = \bigcup_{i=0}^{k} a_i,$$

and $P \setminus \bigcup_{i=0}^{g} a_i$ decomposes into components $P_1$ and $P_2$. Let us number the contours so that $q \in a_0$. Set $b_i = c_i \cup \tau c_i \cup r_i$, where $c_i \subset P_1$ connects $a_0$ with $a_i$, and $r_i \subset a_i$ connects $p_i = c_i \cap a_i$ with $\tau p_i$. The case $(g, k, 1)$ is treated similarly. $\qquad \square$

The definitions immediately imply

LEMMA 8.2. *Let* $\{a_i, b_i \ (i = 1, \ldots, g)\}$ *be a real basis of an algebraic curve* $(P, \tau)$ *of type* $(g, k, \varepsilon)$. *Then* $\bar{h}_j = h_j$ *for* $j \leqslant k - 1$, $\bar{h}_j = h_j - \ell_j$ *for* $\varepsilon = 0, j = k, \ldots, g$, *and* $\bar{h}_j = h_{j+m}$ *for* $\varepsilon = 1, j = k, \ldots, k + m - 1$.

In the rest of this section we assume that the homology basis is real.

**3.** Let $(P, \tau)$ be a real algebraic curve of type $(g, k, \varepsilon)$ and let $J = J(P)$. Consider the involution $\tilde{\tau} \colon \mathbb{C}^g \to \mathbb{C}^g$ which is given, in the basis $(\ell_i, h_i)$ $(i = 1, \ldots, g)$ of $\mathbb{R}^{2g} = \mathbb{C}^g$, by the linear mapping $\ell_j \mapsto \ell_j, h_j \mapsto -h_j$ if $j \leqslant k - 1$ or $\varepsilon = 0$; $\ell_j \mapsto \ell_{j+m}, h_j \mapsto -h_{j+m}$ for $\varepsilon = 1$, $j = k, \ldots, k + m - 1$; and $\ell_j \mapsto \ell_{j-m}, h_j \mapsto -h_{j-m}$ for $\varepsilon = 1, j = k + m, \ldots, g$. By Lemma 8.2, the mapping $\tilde{\tau}$ induces an involution $\tau_\mathbb{R} \colon J \to J$. The same lemma implies that the Abel mapping $A_q$ identifies $\tau_\mathbb{R}$ with the involution $S_g \to S_g$ taking a divisor $D \in S_g$ to $\tau D$.

The fixed points of $\tau_\mathbb{R}$ are called the *real points of the Jacobian of* $(P, \tau)$. They form the *real part* $J_\mathbb{R}$ of the Jacobian.

THEOREM 8.1. *The real part of the Jacobian of a real algebraic curve* $(P, \tau)$ *of type* $(g, k, \varepsilon)$, *where* $k > 0$, *decomposes into* $2^{k-1}$ *real tori of the form*

$$\Phi(T_\mathbb{R} + \delta),$$

*where*

$$\delta = \frac{1}{2} \sum_{j=1}^{k-1} \delta_j h_j, \quad \delta_j \in \{0, 1\},$$

*and* $T_\mathbb{R} = i\mathbb{R}^g$ *if* $\varepsilon = 0$, *and* $T_\mathbb{R} = \{(x_1, \ldots, x_g) \in \mathbb{C}^g \mid x_j \in i\mathbb{R}$ *for* $j \leqslant k - 1$, $\bar{x}_k = -x_{j+m}$ *for* $k \leqslant j \leqslant k + m\}$ *if* $\varepsilon = 1$. *Such a torus is nonsingular if and only if* $\varepsilon = 1$, $k = g + 1$, $\delta_1 = \cdots = \delta_g = 1$.

Proof. The equations for the real part result from a direct calculation. If $p \in P$, then

$$\left( \int_q^p \xi_1 + \int_q^{\tau p} \xi_1, \ldots, \int_q^p \xi_g + \int_q^{\tau p} \xi_g \right) \in T_\mathbb{R}.$$

If $p \in a_j$, then

$$\left( \int_q^p \xi_1, \ldots, \int_q^p \xi_g \right) = \frac{1}{2} h_j.$$

Thus, $x \in \Phi(T_{\mathbb{R}} + \delta)$ if and only if $x = A_q(D)$, where $D \in R_\delta = \{D \in S_g \mid \tau D = D$, the parity of the degree of the divisor $D \cap a_j$ is $\delta_i\}$.

On the other hand, $R_\delta \cap S_{g-1} = \varnothing$ if and only if

$$\sum_{i=1}^{k-1} \delta_i > g - 1,$$

i.e., $k = g + 1$, $\delta_1 = \cdots = \delta_g = 1$.    □

**4.** Together with the involution $\tau_{\mathbb{R}}$, consider the involution $\tau_{\mathbb{I}} = -\tau_{\mathbb{R}} : J \to J$. The fixed points of this involution form the *imaginary part* $J_{\mathbb{I}}$ of the Jacobian.

THEOREM 8.2. *The imaginary part of the Jacobian of a real algebraic curve* $(P, \tau)$ *of type* $(g, k, \varepsilon)$, *where* $k > 0$, *decomposes into* $2^{k-1}$ *real tori of the form* $\Phi(T_{\mathbb{I}} + \delta)$, *where* $\delta = \pi i (\delta_1, \ldots, \delta_{k-1}), \delta_i \in \{0, 1\}$, *and* $T_{\mathbb{I}} = \mathbb{R}^g$ *if* $\varepsilon = 0$, *and* $T_{\mathbb{I}} = \{(x_1, \ldots, x_g) \in \mathbb{C}^g \mid x_j \in \mathbb{R}$ *for* $j \leqslant k - 1$, $\bar{x}_j = -x_{j+m}$ *for* $k \leqslant j \leqslant k + m\}$ *if* $\varepsilon = 1$. *For* $\varepsilon = 0$, *all the tori are singular. For* $\varepsilon = 1$, *exactly one torus is nonsingular, the one corresponding to* $\delta_1 = \delta_2 = \cdots = \delta_{k-1} = 1$.

Proof. The equations for the imaginary part result from a direct calculation. Consider the set $I = \{D \in S_g \mid D + \tau D = $ (the divisor of zeros of a meromorphic differential holomorphic outside $q$ and having a pole of order either 0 or 2 at $q$)$\}$. Then, since $\tau_{\mathbb{R}} K_q = K_q$, we have $A_q(I) = J_{\mathbb{I}}$.

By definition, a meromorphic differential $\xi_D$ is associated to a divisor $D \in I$. Let $A_1 \cup A_2$ be an arbitrary disjoint partition of the set of ovals $A = (a_0, \ldots, a_{k-1})$. Denote by $I_{A_1, A_2} = I_{A_2, A_1} \subset I$ the subset of $D \in I$ such that either the differential $\xi_D$ or the differential $-\xi_D$ is nonnegative on the ovals in $A_1$ and nonpositive on the ovals in $A_2$. The zeros and poles of $\xi_D$ on the ovals are of even orders, whence $I = \bigcup I_{A_1, A_2}$.

By Theorem 6.1, for any partition $A = A_1 \cup A_2$, where $A_1 \neq \varnothing$, $A_2 \neq \varnothing$, there is a holomorphic real differential $\xi$ which is nonnegative on $A_1$ and nonpositive on $A_2$. By adding the differential $\lambda \xi$ to an arbitrary differential $\xi_D$, $D \in I$, it is easy to prove that $I_{A_1, A_2} \neq \varnothing$. Hence $I = \bigcup I_{A_1, A_2}$ consists of at least $2^{k-1}$ connected components. But we have proved already that the set $J_{\mathbb{I}} = A_q(I) - K_q$ consists of $2^{k-1}$ connected components. Therefore, each set $I_{A_1, A_2}$ is connected. If $A_1 \neq \varnothing$ and $A_2 \neq \varnothing$ or $\varepsilon = 0$, then, by Theorem 6.1, there is $D \in I_{A_1, A_2}$ such that $\xi_D$ is holomorphic. In this case $q \in D$,

$$A_q(D) = A_q(D \setminus q) \in A_q(S_{g-1}),$$

and therefore, the connected component $I_{A_1, A_2}$ is singular. If $A_1 = \varnothing$ or $A_2 = \varnothing$, then the assumption $A_q(D) \subset A_q(S_{g-1})$ means that the differential $\xi_D$ is holomorphic and has the same sign on all ovals. For $\varepsilon = 1$, this is impossible, since $\sum_{i=0}^{k-1} a_i = 0$. Therefore, for $\varepsilon = 1$ the connected component $A_q(I_{\varnothing,A})$ is nonsingular.

Let us find the vector $\delta$ to which this connected component corresponds. First let $k = g + 1$ and let $(P, \tau)$ be a hyperelliptic curve. Then the imaginary part of the Jacobian of $(P, \tau)$ coincides with the real part of the Jacobian of the curve $(P, \alpha\tau)$, where $\alpha \colon P \to P$ is the hyperelliptic involution. We assume that $q \in P^\tau \cap P^{\alpha\tau}$ is a fixed point of this involution. Theorem 8.1 implies that the nonsingular imaginary torus of the Jacobian of $(P, \tau)$ (or, what is the same, the nonsingular real torus of the Jacobian of the curve $(P, \alpha\tau)$) corresponds to the vector $\delta = \pi i(1, \ldots, 1)$. This vector does not vary under a continuous deformation of $(P, \tau)$. Because of the connectedness of $M_{g,g+1,1}$ (Theorem 2.1), the same vector corresponds to the nonsingular torus of the imaginary part of the Jacobian for any $M$-curve.

The case $k < g + 1$ can be reduced to the case $k = g + 1$ in the following way. Consider a simple contour $a$ such that $a \cup \tau a$ cuts from $\tilde{P}$ a surface of genus $k - 1$ with 2 holes. Consider a continuous deformation of $(P, \tau)$ contracting $a$ into a point. In the process of deformation, the vector corresponding to the nonsingular torus of the imaginary part of the Jacobian remains constant. In the limit, it gives the vector corresponding to the $M$-curve, that is, $\pi i(1, \ldots, 1)$. $\qquad\square$

**Remark.** The number of real and imaginary tori of the Jacobian was first found in [**16**]. The number of real and imaginary tori was computed for the case $\varepsilon = 1$ in [**26**], and for the case $\varepsilon = 0$ in [**21**], [**65**]. Another proof is given in [**92**].

## 9. Prymians of real algebraic curves

**1.** In the classical algebraic geometry (see, e.g. [**26**]), algebraic varieties of another kind, close to Jacobians, but different from them, are associated to curves with automorphisms. We shall consider only the simplest class of such varieties, which has, however, important applications.

Let $P$ be a compact Riemann surface of genus $2g$ and let $\alpha \colon P \to P$ be a holomorphic involution with two fixed points, $q_1$ and $q_2$. A symplectic basis $\{a_i, b_i \ (i = 1, \ldots, 2g)\}$ is said to be *symmetric* if $\alpha a_i = -a_{i+g}$, $\alpha b_i = -b_{i+g}$ $(i = 1, \ldots, g)$. The mapping of divisors $D \mapsto \alpha(D)$ induces an involution $\alpha^* \colon S_g \to S_g$. The Abel mapping $A_{q_1}$ pushes this involution to $J = J(P)$ thus inducing an involution $\alpha^* \colon J \to J$. The subset

$$\mathrm{Pr} = \mathrm{Pr}(P, \alpha) = \{x \in J \,|\, \alpha^* x = -x\}$$

is called the *Prymian* of the surface with involution $(P, \alpha)$. A Prymian is isomorphic to the torus $\mathbb{C}^g / G$, where $G$ is a lattice generated by the vectors $\ell_i$ and the columns $\xi_i$ of the matrix

$$A_{ij} = \int_{b_i} \xi_j + \xi_{j+g} \quad (i, j = 1, \ldots, g),$$

in the notation of Section 7.

**2.**    A real curve with an involution $(P, \tau_1, \alpha)$ is a compact Riemann surface $P$ of genus $2g$ endowed with two involutions: an antiholomorphic involution $\tau_1$ and a holomorphic involution $\alpha$ having exactly two fixed points $q_1, q_2$, with the condition $\tau_1 q_1 = q_2$ satisfied. Let us set $\tau_2 = \tau_1 \alpha$. Suppose there are $r_i$ ovals of the involution $\tau_i$ invariant under $\alpha$ and $2t_i$ ovals transposed in pairs by $\alpha$. Then

$$(\tilde{P}, \tilde{\tau}) = (P/\langle \alpha \rangle, \tau_i/\langle \alpha \rangle)$$

is a real algebraic curve of type $(g, k, \varepsilon)$, where $k = t_1 + r_1 + t_2 + r_2$, and the preimage of $\tilde{P}^{\tilde{\tau}}$ coincides with $P^{\tau_1} \cup P^{\tau_2}$. It cuts $P$ into two parts if and only if $\varepsilon = 1$. We call the set $(g, \varepsilon, t_1, r_1, t_2, r_2)$ the *type of the real algebraic curve with involution* $(P, \tau_1, \alpha)$.

EXAMPLE 9.1. Let $(\tilde{P}, \tilde{\tau})$ be a real curve of type $(g, k, 1)$ and let $k = t_1 + r_1 + t_2 + r_2$, where $r_1 + r_2 = 1 \pmod 2$. Consider a connected component $\tilde{P}_1$ of $\tilde{P} \setminus \tilde{P}^\tau$ and let $\phi_1 \colon P_1 \to \tilde{P}_1$ be the two-sheeted ramified covering with a single ramification point $q_1 \in P_1$, which is two-sheeted on the $r_1 + r_2$ contours $c_1, \ldots, c_{r_1+r_2} \in \varphi P_1$ and one-sheeted on the other boundary contours $c_{r_1+r_2+1}, \ldots, c_{\hat{k}}$, where $\hat{k} = r_1 + r_2 + 2t_1 + 2t_2$. Using the construction in Example 1.1, we construct a real algebraic curve $(\hat{P}, \hat{\tau})$ such that $\hat{P}^{\hat{\tau}} = \bigcup_{i=1}^{\hat{k}} c_i$ cuts $\hat{P}$ into $P_1$ and $P_2 = \hat{\tau} P_1$. The covering $\phi_1$ induces the two-sheeted covering $\hat{\phi} \colon \hat{P} \to \tilde{P}$, where $\hat{\phi}\hat{\tau} = \tilde{\tau}\hat{\phi}$. Let $\alpha \colon \hat{P} \to \hat{P}$ be the involution transposing the sheets of the covering. It commutes with $\hat{\tau}$ and has exactly two fixed points $q_1$ and $q_2 = \hat{\tau} q_1$. Let us cut the surface $\hat{P}$ along the contours $c_{r_1+1}, \ldots, c_{r_1+r_2}$ and $c_{r_1+r_2+2t_1+1}, \ldots, c_{r_1+r_2+2t_1+2t_2}$ and glue the boundary contours along the mapping $\alpha\hat{\tau}$. The involution $\hat{\tau}$ induces on the resulting surface $P$ an involution $\tau_1 \colon P \to P$ commuting with $\alpha$. Let us set $\tau_2 = \alpha\tau_1$. It is easy to see that $(P, \tau_1, \alpha)$ is a real curve with involution, of type $(g, 1, t_1, r_1, t_2, r_2)$.

It is also easy to see that the following statement is true.

LEMMA 9.1. *The construction of Example 9.1 produces all curves with involution, of type* $(g, 1, t_1, r_1, t_2, r_2)$.

EXAMPLE 9.2. Let $(\tilde{P}, \tilde{\tau})$ be a real curve of type $(g, k, 0)$ and let $k = t_1 + r_1 + t_2 + r_2$, where $r_1 + r_2 = 1 \pmod 2$. Using Lemma 1.2, choose a set of pairwise disjoint contours $\tilde{c}_1, \ldots, \tilde{c}_{g+1}$ such that $\tilde{\tau}\tilde{c}_i = \tilde{c}_i$ and $\tilde{P}^{\tilde{\tau}} = \bigcup_{i=1}^k \tilde{c}_i$. Consider a connected component $\tilde{P}_1$ of the set $\tilde{P} \setminus \bigcup_{i=1}^{g+1} \tilde{c}_i$ and the two-sheeted covering $\phi_1 \colon P_1 \to \tilde{P}_1$ with the unique ramification point $q_1 \in P_1$ that is two-sheeted over the contours $c_1, \ldots, c_{r_1+r_2}$ and one-sheeted over the remaining contours $c_{r_1+r_2+1}, \ldots, c_v$. Using the construction of Example 1.2, we construct a real algebraic curve $(\hat{P}, \hat{\tau})$ such that $\hat{P} \setminus \bigcup_{i=1}^v c_i$ cuts $\hat{P}$ into $P_1$ and $P_2 = \hat{\tau} P_1$ and $\hat{P}^{\hat{\tau}} = \bigcup_{i=1}^{\hat{k}} c_i$, where $\hat{k} = r_1 + r_2 + 2t_1 + 2t_2$. Repeating word for word the cutting and gluing process in Example 9.1, we obtain a real curve with involution $(P, \tau_1, \alpha)$, of type $(g, 0, t_1, r_1, t_2, r_2)$.

LEMMA 9.2 ([66], [12]). *The construction of Example* 9.2 *produces all real curves with involution, of type* $(g, 0, t_1, r_1, t_2, r_2)$.

**3.** Let $(P, \tau_1, \alpha)$ be a real curve with involution, of type $(g, \varepsilon, t_1, r_1, t_2, r_2)$. The intersection of the Prymian $\mathrm{Pr} = \mathrm{Pr}(P, \alpha) \subset J(P) = J$ with the real part of the Jacobian of $(P, \tau_1)$ is called the *real part of the Prymian* of the real curve with involution $(P, \tau_1, \alpha)$. Its connected components are called the *real tori* of the Prymian of $(P, \tau_1, \alpha)$. They form the set of fixed points of the involution $(\tau_1)_{\mathbb{R}}|_{\mathrm{Pr}} \colon \mathrm{Pr} \to \mathrm{Pr}$.

THEOREM 9.1. *The real part of the Prymian of a real curve with involution* $(P, \tau_1, \alpha)$ *of type* $(g, \varepsilon, t_1, r_1, t_2, r_2)$, *where* $k = t_1 + r_1 + t_2 + r_2 > 0$, *decomposes into* $2^{k-1}$ *real tori of dimension* $g$.

Proof. Let us choose a symmetric basis $\Delta = \{a_i, b_i \ (i = 1, \ldots, g)\}$ of the pair $(P, \alpha)$ in such a way that the projection of the cycles $\{a_i, b_i \ (i = 1, \ldots, g)\}$ is a real basis $\tilde{\Delta}$ of the real curve $(\tilde{P}, \tilde{\tau}) = (P/\langle \alpha \rangle, \tau_1/\langle \alpha \rangle)$ of type $(g, k, \varepsilon)$. Let $\{\ell_i, d_i\}$ be the generators of the lattice of the Prymian $\mathrm{Pr}$ of the real curve with involution $(P, \tau_1, \alpha)$ corresponding to the basis and let $\{\tilde{\ell}_i, \tilde{h}_i\}$ be the generators of the lattice of the Jacobian $\tilde{J}$ of the real curve $(\tilde{P}, \tilde{\tau})$. In these bases, the involutions $(\tau_1)_{\mathbb{R}}|_{\mathrm{Pr}} \colon \mathrm{Pr} \to \mathrm{Pr}$ and $\tilde{\tau}_{\mathbb{R}} \colon \tilde{J} \to \tilde{J}$ are described by coinciding formulas and, therefore, they have the same number of fixed tori.  □

**4.** Let $(P, \tau_1, \alpha)$ be a real curve with involution, of type $(g, \varepsilon, t_1, r_1, t_2, r_2)$. Let us number the ovals $a_1^j, \ldots, a_{2t_j + r_j}^j$ of $\tau_j$ so that $\alpha a_i^j = a_{t_j + i}^j$ for $i \leqslant t_j$. The set $\Omega$ consists of divisors $D \subset P$ of degree $g$ such that $\tau_1 D = D$ and $\alpha D + D$ is the divisor of zeros of a meromorphic differential $\xi_D$ holomorphic outside the fixed points $q_1, q_2$ of the involution $\alpha$ and having poles of order 0 or 1 at these points. We say that $\xi_D$ is positive definite on an oval $a = a_i^1$, where $i > 2t_1$, if either $\xi_D$ is nonnegative on $a$, or there is a point $p \in a \cap D$ that, together with the point $\alpha p$, cuts the contour $a$ into two open arcs so that the arc of positiveness of $\xi_D$ contains evenly many points of the divisor $D$ in a neighborhood of $p$. Otherwise, we say that $\xi_D$ is negative definite on $a$. We say that $\xi_D$ is positive (respectively, negative) definite on an oval $a_i^2$ if it is nonnegative (respectively, nonpositive) on it, as a real differential of the curve $(P, \tau_2)$.

Let us split the set

$$a_{2t_1+1}^1, \ldots, a_{2t_1+r_1}^1, a_1^2, \ldots, a_{2t_2+r_2}^2$$

into the subsets $A_1$ and $A_2$. Let

$$\delta = (\delta_1, \ldots, \delta_{t_1}) \in \mathbb{Z}_2^{t_1}.$$

Denote by $\Omega(\delta, A_1, A_2)$ the subset of $\Omega$ consisting of divisors $D \in \Omega$ such that either the differential $\xi_D$ or the differential $-\xi_D$ is positive definite on $A_1$, negative definite on $A_2$, and the parity of the degree of the differential $D \cap a_i^1$ for $i \leqslant t_1$ coincides with the parity of $\delta_i$.

LEMMA 9.3. *Each of the sets* $\Omega(\delta, A_1, A_2)$ *is nonempty.*

Proof. Let us prove first that on each real algebraic curve $(\tilde{P}, \tilde{\tau})$ with the ovals $c_1, \ldots, c_k$, where $k = k_+ + k_- + k_0$, and for each pair of points $q_1 \neq q_2$, where $q_2 = \tilde{\tau} q_1$, there is a meromorphic real differential $\xi$ that is holomorphic outside $q_1, q_2$, has a pole of order at most 1 at these points, is nonnegative on $c_i$ for $i \leqslant k_+$, nonpositive on $c_i$ for $k_+ < i \leqslant k_+ + k_-$, and has zeros on $c_i$ for $i > k_+ + k_-$. To this end, encircle the points $q_1$ and $q_2$ by disjoint neighborhoods $U_i \supset q_i$ such that $\tilde{\tau} U_1 = U_2$, and glue the boundaries of the surface $\tilde{P} \setminus (U_1 \cup U_2)$ by means of the involution $\tilde{\tau}$. Under this process, the boundary is taken to an oval $c_0$ of the new real algebraic curve $(P', \tau')$. Applying to this curve Theorem 6.1, we find a real differential $\xi'$ with the desired behavior on the ovals $c_1, \ldots, c_k$. Now the desired differential on $(\tilde{P}, \tilde{\tau})$ can be obtained by contracting the oval $c_0$.

Applying this result to the real curve $(\tilde{P}, \tilde{\tau}) = (P/\langle \alpha \rangle, \tau_1/\langle \alpha \rangle)$, we find a meromorphic differential that is holomorphic outside the images $\tilde{q}_1, \tilde{q}_2$ of the points $q_1, q_2$, has at most simple poles at these points, is nonnegative on the images of ovals in $A_1$, nonpositive on the images of poles in $A_2$, and has zeros on the other ovals. Its preimage $\xi$ on $P$ is a meromorphic differential holomorphic outside $q_1$ and $q_2$, having at most simple poles at these points, positive definite on the ovals in $A_1$ and negative definite on the ovals in $A_2$. Its divisor of zeros, when intersected with $a_i^1 \cup \alpha a_i^1$ $(i \leqslant t_1)$, has a positive degree divisible by 4 and is symmetric with respect to $\alpha$. Therefore, there is a divisor $D \in \Omega$ such that $\xi_D = \xi$ and the parity of the degree of $D \cap a_i^1$ coincides with the parity of $\delta_i$ for $i \leqslant t_1$. $\square$

THEOREM 9.2 ([**65**], [**78**]). *Let* $(P, \tau, \alpha)$ *be a real curve with involution, of type* $(g, \varepsilon, t_1, r_1, t_2, r_2)$, *where* $k = t_1 + r_1 + t_2 + r_2 > 0$. *Then*
  (1) *for* $\varepsilon = 0$ *all real tori of its Prymian are singular;*
  (2) *for* $\varepsilon = 1$ *the Prymian contains at most one nonsingular real torus;*
  (3) *for* $\varepsilon = 1$, $k = g + 1$ *the Prymian always contains a real nonsingular torus;*
  (4) *for* $\varepsilon = 1$, $t_1 + r_1 \leqslant \frac{k}{2}$ *there is a curve* $(P, \tau_1, \alpha)$ *of type* $(g, \varepsilon, t_1, r_1, t_2, r_2)$ *such that its Prymian contains a nonsingular real torus.*

Proof. It is easy to see that

$$\Omega(\delta, A_1, A_2) \cap \Omega(\delta', A_1', A_2') \neq \varnothing$$

if and only if $\delta' = \delta$, $A_1' = A_2$, $A_2' = A_1$, and if this assumption is satisfied, then the two sets coincide. Hence the number of disjoint sets $\Omega(\delta, A_1, A_2)$ is $2^{k-1}$. On the other hand, the real part of the Prymian of a curve with involution $(P, \tau_1, \alpha)$ coincides with $\bigcup A_{q_1}(\Omega(\delta, A_1, A_2)) - K_{q_1}$ and, by Theorem 9.1, consists of $2^{k-1}$ connected components. Hence, by Lemma 9.3, each real torus of the Prymian has the form

$$A_{q_1}(\Omega(\delta, A_1, A_2)) - K_{q_1}.$$

A torus is singular if and only if there is $D \in \Omega(\delta, A_1, A_2)$ such that $\alpha D + D$ is the divisor of zeros of a holomorphic differential on $P$.

(1) Let $\varepsilon = 0$ and let $T = A_{q_1}(\Omega(\delta, A_1, A_2)) - K_{q_1}$ be an arbitrary real torus of the Prymian. Let $\tilde{A}_i$ be the image of the set $A_i$ on the real curve $(\tilde{P}, \tilde{\tau}) = (P/\langle\alpha\rangle, \tau_1/\langle\alpha\rangle)$. By Theorem 6.1, there is a holomorphic real differential $\tilde{\xi}$ on $(\tilde{P}, \tilde{\tau})$ nonnegative on $\tilde{A}_1$, nonpositive on $\tilde{A}_2$ and having zeros on other ovals. Its preimage $\xi$ on $P$ is a holomorphic differential that is positive definite on the ovals in $A_1$ and negative definite on the ovals in $A_2$. Its divisor of zeros, when intersected with $a_i^1 \cup \alpha a_i^1$ ($i \leqslant t_1$), has a positive degree divisible by 4 and is symmetric with respect to $\alpha$. Therefore, there is a divisor $D \in \Omega(\delta, A_1, A_2)$ such that $\xi_D = \xi$, and $T$ is a singular torus.

(2) Let $\varepsilon = 1$ and let $T = A_{q_1}(\Omega(\delta, A_1, A_2)) - K_{q_1}$ be an arbitrary torus distinct from $A_{q_1}(\Omega(\delta, A, \varnothing)) - K_{q_1}$, where $\delta = (1, \dots, 1)$. Then, repeating the argument in the case $\varepsilon = 0$, we deduce that $T$ is a singular torus.

(3) Let $\varepsilon = 1$ and let $k = g + 1$. For $\delta = (1, \dots, 1)$ the real torus $A_{q_1}(\Omega(\delta, A_1, \varnothing)) - K_{q_1}$ is nonsingular. Indeed, otherwise there would exist a real holomorphic differential $\xi$ on $P$ positive definite on all the ovals of the involutions $\tau_1$ and $\tau_2$, where it has no zeros, and such that $\alpha\xi = \xi$. This differential would induce a holomorphic real differential $\tilde{\xi}$ on the $M$-curve $(\tilde{P}, \tilde{\tau}) = (P/\langle\alpha\rangle, \tau_1/\langle\alpha\rangle)$, which is positive on all ovals where it has no zeros. But Theorem 6.2 asserts that there are no such differentials.

(4) Let $\varepsilon = 1$ and $t_1 + k_1 \leqslant k/2$ and let $T$ be a real torus of the form $A_{q_1}(\Omega(\delta, A_1, \varnothing)) - K_{q_1}$, where $\delta = (1, \dots, 1)$. If the torus $T$ is singular, then, repeating the argument used in the case $k = g + 1$, we obtain a differential $\tilde{\xi}$ on the real curve $(\tilde{P}, \tilde{\tau}) = (P/\langle\alpha\rangle, \tau_1/\langle\alpha\rangle)$, which is nonnegative on the $t_2 + k_2 > k/2$ images of the ovals in $A_1$ and either has zeros or is positive on the other ovals of the curve. Example 9.1 shows that any real curve can appear as a curve $(\tilde{P}, \tilde{\tau})$, in particular, those constructed in Theorem 6.3, where there are no such differentials.  $\square$

**Remark.** Under a small deformation of the curve with involution $(P, \tau_1, \alpha)$, a nonsingular torus is transformed into a nonsingular torus. Therefore, curves with involution $(P, \tau_1, \alpha)$ whose Prymian contains a nonsingular real torus form an open set in the space of all curves of a given type.

## 10. Uniformization of real algebraic curves by Schottky groups

**1.** Let $\psi \in \tilde{T}_{\tilde{g}, k}$, $k > 0$, and let $\{A_i, B_i \ (i = 1, \dots, \tilde{g}), \ C_i \ (i = 1, \dots, k)\} = \psi(\gamma_{\tilde{g}, k})$ be the corresponding sequential set of shifts. Set

$$\Delta = \{A_i, B_i \ (i = 1, \dots, \tilde{g}), C_i \ (i = 1, \dots, k - 1)\}.$$

The invariant curves $\ell(\Delta)$ of the set $\Delta$ are shown in Fig. 2.10.1.

According to Sections 1–3 of Chapter 1, for any $D \in \Delta$ there are disks $S_D$ and $S_{D^*}$ centered on $\mathbb{R} \cup \infty$ such that $S_D \cap \ell(\Delta) \subset \ell(D)$, $S_{D^*} \cap \ell(\Delta) \subset \ell(D)$, $D(\ell(\Delta) \setminus \ell(D)) \subset S_D$ and $D^{-1}(\ell(\Delta) \setminus \ell(D)) \subset S_{D^*}$. It is easy to show, using the methods of Sections 1–3 of Chapter 1, that the disks $S_D$ and $S_{D^*}$ can

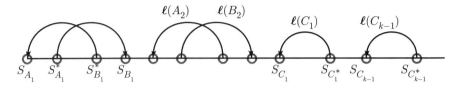

FIGURE 2.10.1

be chosen in such a way that

$$S_{D_1} \cap S_{D_2} = S_{D_1^*} \cap S_{D_2} = S_{D_1^*} \cap S_{D_2^*} = \varnothing \quad \text{for} \quad D_1 \neq D_2$$

and $D(\varphi S_{D^*}) = \varphi S_D$. In this case $\Omega = \mathbb{C} \cup \infty \setminus \bigcup_{D \in \Delta}(S_D \cup S_{D^*})$ is a fundamental domain of the Schottky group $G$ generated by $\Delta$. On the quotient surface $P = \Omega/G$ of genus $g = 2\tilde{g} + k - 1$, the involution $z \mapsto \bar{z}$ induces a separating involution $\tau \colon P \to P$ with $k$ ovals.

In order to show that this construction produces all separating real algebraic curves, it suffices to construct for such a curve $(P, \tau)$ a system of cuts on a connected component $P_1$ of the surface $P \setminus P^\tau$ making $P_1$ into a half of a fundamental Schottky domain. Such a system of cuts was suggested in [9]; it is shown in Fig. 2.10.2.

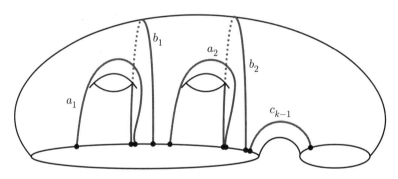

FIGURE 2.10.2

Hence (see [9], [10]) the correspondence $\psi \mapsto (P, \tau)$ determines a mapping

$$\Psi_k \colon T_{\tilde{g},k} \to M_{g,k,1},$$

where $\Psi_k(T_{\tilde{g},k}) = M_{g,k,1}$.

A similar description of nonseparating curves with real points can be obtained if we replace the system $\Delta$ of generators of the Schottky group $G$ by the set

$$\Delta^* = \{A_i, B_i \ (i = 1, \ldots, \tilde{g}), C_i \ (i = 1, \ldots, \tilde{k}), C_i^* \ (i = \tilde{k}+1, \ldots, k-1)\},$$

where $C_i^*$ is obtained from

$$C_i(z) = \frac{(\lambda_i \alpha_i - \beta_i)z - (1 - \lambda)\alpha_i \beta_i}{(\lambda_i - 1)z + (\alpha_i - \lambda_i \beta_i)}$$

by substituting $-\lambda_i$ for $\lambda_i$. The system $\Delta^*$ generates the Schottky group $G^*$. On the quotient surface $P^* = \Omega/G^*$, the involution $z \mapsto \bar{z}$ induces a nonseparating involution $\tau^*\colon P^* \to P^*$ with $\tilde{k}+1$ ovals. Thus the correspondence $\psi \mapsto (P^*, \tau^*)$ determines a mapping

$$\Psi_{\tilde{k}+1}\colon T_{\tilde{g},k} \to M_{g,\tilde{k}+1,0}.$$

The identity

$$\Psi_{\tilde{k}+1}(T_{\tilde{g},k}) = M_{g,\tilde{k}+1,0}$$

can be proved in the same way as in the case of separating curves. The only thing we need to do is to complete the set of ovals of the curve $(P^*, \tau^*)$ to a system of pairwise disjoint invariant contours $c_1, \ldots, c_k$ so that the surface $P^* \setminus \bigcup_{i=1}^{k} c_i$ decomposes into two connected components. Hence each moduli space $M_{g,\tilde{k},\varepsilon}$ admits a representation

$$M_{g,\tilde{k},\varepsilon} = \Psi_{\tilde{k}}(T_{\tilde{g},k}).$$

Because of the theorem

$$T_{\tilde{g},k} \cong \mathbb{R}^{6\tilde{g}+3k-6}$$

(see [**55**], [**85**]), this gives another proof of Corollary 2.1:

$$M_{g,k,\varepsilon} \cong \mathbb{R}^{6g-6}/\operatorname{Mod}_{g,k,\varepsilon}.$$

**2.** The Schottky uniformization allows one to solve the Schottky problem for real algebraic curves, i.e., to find matrices $B_{ij}$ introduced in Section 8.

Let us describe the matrix corresponding to the system of generators

$$\tilde{\Delta} = \{A_i, B_i \ (i = 1, \ldots, \tilde{g}), \tilde{C}_i \ (i = 1, \ldots, k-1)\} = \{\tilde{D}_i \ (i = 1, \ldots, 2\tilde{g}+k-1)\}$$

of a Schottky group $G$ of the above-described form. Let

$$D_i(z) = \frac{(\lambda_i\alpha_i - \beta_i)z - (1-\lambda)\alpha_i\beta_i}{(\lambda_i - 1)z + (\alpha_i - \lambda_i\beta_i)}.$$

Denote by $G_{mn}$ the subset of $G$ consisting of the elements of the form

$$D = D_{i_1}^{j_1} \cdots D_{i_k}^{j_k},$$

where $j_\ell \neq 0, i_1 \neq m, i_k \neq n$. Set

$$\{z_1, z_2, z_3, z_4\} = \frac{(z_1 - z_2)(z_2 - z_4)}{(z_1 - z_4)(z_2 - z_3)}.$$

Then, according to [**5**] and [**11**], the Jacobi matrix $(B_{nm})$ of the algebraic curve $\Omega/\tilde{g}$ corresponding to the generators $\tilde{\Delta}$ is given by the convergent series

(10) $$B_{nn} = \ln\lambda_n + \sum_{D \in G_{nn}} \ln\{\alpha_n, \beta_n, D\alpha_n, D\beta_n\}$$

and

(11) $$B_{nm} = \sum_{D \in G_{mn}} \ln\{\alpha_m, \beta_m, D\alpha_n, D\beta_n\} \text{ for } m \neq n.$$

Hence Eqs. (10), (11) together with the explicit description of the space $T_{\tilde{g},k}$ by means of the coordinates $\{\alpha_i, \beta_i, \lambda_i \ (i = 1, \ldots, 2\tilde{g} + k)\}$ (Sections 3–4 of Chapter 1) allow one to find the Jacobians of algebraic real curves and to find, using the formulas in Section 8, their real and imaginary tori.

A modification of this approach allows one to describe the Prymians of real algebraic curves [65].

## 11. Moduli spaces of spinor bundles of rank 1 on real algebraic curves

**1.** We recall that Fuchsian groups uniformizing Riemann surfaces of genus 0 have the form $\psi(\gamma_{0,g+1})$, where $\gamma_{0,g+1}$ is the group generated by $c_1, \ldots, c_{g+1}$ modulo the defining relation $c_1 \cdots c_{g+1} = 1$, and $\psi \colon \gamma_{0,g+1} \to \operatorname{Aut}(\Lambda)$ is a monomorphism belonging to the set $\tilde{T}_{0,g+1}$ (Sections 1, 2 of Chapter 1). Associated to such a monomorphism is the group $\Gamma_\psi^k$ $(k \leqslant g)$ generated by $\psi(\gamma_{0,g+1})$ and the elements

$$\hat{C}_i = \begin{cases} \overline{C}_i & \text{for } i \leqslant k, \\ \tilde{C}_i & \text{for } i > k, \end{cases}$$

where $C_i = \psi(c_i)$. Set $D_i = \hat{C}_{g+1}\hat{C}_i \ (i = 1, \ldots, g)$. The natural isomorphism $\Gamma_\psi^k \to \pi_1(P, p)$, where $P = \Lambda/\Gamma_\psi^k$, takes $\{C_i, D_i \ (i = 1, \ldots, g)\}$ to the generators $\{c_i, d_i \ (i = 1, \ldots, g)\}$ of the group $\pi_1(P, p)$, subject to the defining relation

$$\prod_{i=1}^{g} c_i \prod_{i=g}^{1} d_i c_i^{-1} d_i^{-1} = 1.$$

LEMMA 11.1. *Let $\tilde{\Gamma}^*$ be a lifting of a real Fuchsian group $\tilde{\Gamma}$, where $[\tilde{\Gamma}] = (P, \tau)$ is a real algebraic curve of type $(g, k, 0)$. Then there is a monomorphism $\psi \in \tilde{T}_{0,g+1}$ such that $\tilde{\Gamma} = \Gamma_\psi^k$ and $\omega_{\tilde{\Gamma}^*}(d_i) = \omega_{\tilde{\Gamma}^*}(d_j)$ for all $i, j \leqslant g$.*

Proof. Consider a set of contours $c_1, \ldots, c_{g+1}$ possessing the properties described in Theorem 4.2. Let $P_1$ be a sphere with $g + 1$ holes whose boundary $\partial P_1$ consists of the contours $\tilde{c}_1, \ldots, \tilde{c}_{g+1}$ endowed with the orientation induced by $\tilde{\Gamma}^*$. Consider the standard system of generators $(c_1, \ldots, c_{g+1})$ of $\pi(P_1, p)$ associated to this system of contours. Let us identify the $c_i$ with the standard generators of the group $\gamma_{0,g+1}$. Then the natural isomorphism $\pi_1(P, p) \to \Gamma$ induces $\psi \in \tilde{T}_{0,g+1}$. It is easy to see that $\Gamma_\psi^k = \tilde{\Gamma}$.

Now let us find $\omega_{\tilde{\Gamma}^*}(d_i)$. Set $\hat{C}_i^* = J^{-1}(\hat{C}_i) \cap \tilde{\Gamma}^*$. Replacing $\tilde{\Gamma}$ by a conjugate group, if necessary, we may assume that

$$\hat{C}_{g+1} = \sigma_{g+1}\begin{pmatrix} -\mu_{g+1} & 0 \\ 0 & \mu_{g+1}^{-1} \end{pmatrix},$$

where $\mu_{g+1} > 0$, $\sigma_{g+1} = \pm 1$. The shifts $C_1, \ldots, C_{g+1}$ form a sequential set (see Section 2), and therefore, the configuration of their invariant lines $\ell(C_i)$ is as shown in Fig. 2.11.1.

Hence

$$\overline{C}_i = F_i \overline{C}_{g+1} F_i^{-1},$$

where

$$F_i = \frac{(\lambda_i \alpha_i + \alpha_i)z - (1 - \lambda_i)\alpha_i^2}{(1 - \lambda_i)z + (\alpha_i + \lambda_i \alpha_i)} = \frac{\alpha_i(\lambda_i + 1)z + (\lambda_i - 1)\alpha_i^2}{(1 - \lambda_i)z + \alpha_i(\lambda_i + 1)}.$$

We set

$$F_i^* = \begin{pmatrix} \alpha_i(\lambda_i + 1) & (\lambda_i - 1) \\ (1 - \lambda_i) & \alpha_i(\lambda_i + 1) \end{pmatrix}.$$

Then

$$\hat{C}_i^* = \sigma_i F_i^* \begin{pmatrix} -\mu_i & 0 \\ 0 & \mu_i^{-1} \end{pmatrix} (F_i^*)^{-1},$$

where $\mu_i > 0$, $\sigma_i = \pm 1$. Let us prove that for $i \leqslant g$ we have $\sigma_i = -1$. Indeed, by the construction, the orientation of the contour $c_i$ induced by $\tilde{\Gamma}^*$ coincides with the orientation of the same contour as the boundary of $P_1$. This orientation induces the orientation of the line $\ell(C_i)$ shown in Fig. 2.11.1. The mapping $F^{-1}$ takes this orientation to that of the imaginary axis $I$ given by decreasing of Im $z$ (see Fig. 2.11.1). And this means that $\sigma_i = -1$.

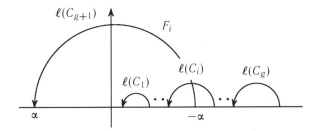

FIGURE 2.11.1

Thus

$$\hat{C}_{g+1}^* \hat{C}_i^* = -\sigma_{g+1} \begin{pmatrix} \mu_{g+1} & 0 \\ 0 & \mu_{g+1}^{-1} \end{pmatrix} F_i^* \begin{pmatrix} \mu_i & 0 \\ 0 & \mu_i^{-1} \end{pmatrix} (F_i^*)^{-1},$$

whence

$$\omega_{\Gamma^*}(d_i) = \text{sign}(\text{Tr}(D_i)) = \text{sign}(\text{Tr}(\hat{C}_{g+1}^* \hat{C}_i^*)) = -\sigma_{g+1},$$

and therefore, $\omega_{\tilde{\Gamma}^*}(d_i)$ is the same for all $i \leqslant g$. □

LEMMA 11.2. *Let $\omega$ be a nonspecial Arf function on a real algebraic curve $(P, \tau)$ of type $(g, k, 0)$. Then there is a standard basis*

$$\{c_i, d_i \ (i = 1, \ldots, g)\} \in H_1(P, \mathbb{Z}_2)$$

*such that $c_1, \ldots, c_g$ are pairwise disjoint invariant contours $\tau(d_i) = d_i + c_{g+1} + \tilde{c}_i$, where*

$$\tilde{c}_i = \begin{cases} 0 & \text{for } i \leqslant k, \\ c_i & \text{for } i > k, \end{cases}$$

*and $\omega(d_i) = \omega(d_j)$ for all $i, j \leqslant g$.*

Proof. By Lemma 2.1, there is a real Fuchsian group $\tilde{\Gamma}$ such that $(P, \tau) = [\tilde{\Gamma}]$. By Lemma 4.2, it has a lifting $\tilde{\Gamma}^*$ such that $\omega_{\tilde{\Gamma}^*} = \omega$. Therefore, the assertion of Lemma 11.2 follows from Lemma 11.1.          □

**2.** Let $(P, \tau)$ be a real algebraic curve. Two Arf functions $\omega_1$ and $\omega_2$ on $(P, \tau)$ are said to be *topologically equivalent* if there is a homeomorphism $\phi \colon P \to P$ such that $\phi\tau = \tau\phi$ and the induced automorphism $\phi \colon H_1(P, \mathbb{Z}_2) \to H_1(P, \mathbb{Z}_2)$ possesses the property $\omega_1 = \omega_2\phi$.

THEOREM 11.1 ([**76**]). *All special Arf functions on an arbitrary curve $(P, \tau)$ are topologically equivalent.*

Proof. By Lemma 3.2, we have $P^\tau = \varnothing$. Therefore, by Lemma 1.2, there is a set of pairwise disjoint invariant contours $c_1, \ldots, c_{g+1}$ such that the complement $P \setminus \bigcup_{i=1}^{g+1} c_i$ consists of two spheres with holes $P_1$ and $P_2$. Connect the contour $c_i$ by a simple segment $\ell_i \subset P_1$ with the contour $c_{g+1}$. Consider the simple closed contour $d_i = \ell_i \cup \tau\ell_i \cup r_i \cup r_{g+1}$, where $r_j \subset c_j$. Let $\omega_1$ and $\omega_2$ be special Arf functions on $(P, \tau)$. By Lemma 3.3, we have $\omega_1(c_i) = \omega_2(c_i) = 0$. For each $i$ such that $\omega_1(d_i) \neq \omega_2(d_i)$ we apply to $P$ the Dehn twisting along the contour $c_i$, i.e., we cut the surface $P$ along $c_i$ and attach it back after twisting by the angle $2\pi$. It is easy to see that the homeomorphism $\phi$ thus obtained commutes with $\tau$. On the other hand, for such $i$ we have

$$\omega_2(\phi(d_i)) = \omega_2(d_i + c_i) = \omega_2(d_i) + \omega(c_i) + 1 = \omega_1(d_i).$$

Hence, $\omega_2\phi = \omega_1$.          □

THEOREM 11.2 ([**76**]). *Let $(P, \tau)$ be a real algebraic curve of type $(g, k, 0)$. Then two nonspecial Arf functions on $(P, \tau)$ are topologically equivalent if and only if they have the same topological type $(g, \delta, k_\alpha)$.*

Proof. Let $\omega_1$ and $\omega_2$ be two nonspecial Arf functions on $(P, \tau)$, of type $(g, \delta, k_\alpha)$. Using Lemma 11.2 we associate to the Arf function $\omega_m$ a standard basis $\{c_i^m, d_i^m \ (i = 1, \ldots, g)\}$, where $c_i^m$ are pairwise disjoint invariant contours and $\omega_m(d_i^m) = \omega_m(d_j^m)$ for all $i, j \leqslant g$. After an appropriate renumbering, one can assume that $c_1^m, \ldots, c_{k_0+k_1}^m$ are ovals and $\omega_m(c_i^m) = 0$ for $i \leqslant k_0$ and $\omega_m(c_i^m) = 1$ for $i > k_0$. By Theorem 3.2, we have $k_0 \equiv g + 1 \pmod 2$ and, therefore,

$$\delta(P, \omega_m) = \sum_{j=1}^{g+1} \omega_m(c_j^m)\omega_m(d_j^m) = \sum_{j=k_0+1}^{g} \omega_m(d_j^m) = \omega_m(d_j^m).$$

Thus, $\omega_1(d_j^1) = \delta = \omega_2(d_j^2)$. By Lemma 1.2, one can add to the set $c_1^m, \ldots, c_g^m$ a new contour $c_{g+1}^m$ thus making it into a complete set of invariant contours. Let

$$P_1^m \cup P_2^m = P \setminus \bigcup_{i=1}^{g+1} c_i^m.$$

Choose $C_{g+1}^m$ in such a way that the homeomorphism $\phi\colon P_1^1 \to P_1^2$ can be extended to a homeomorphism $\phi\colon P^1 \to P^2$ commuting with $\tau$ and taking $\{c_i^1, d_i^1\}$ to $\{c_i^2, d_i^2\}$. Then $\omega_1 = \omega_2\phi$.                                    $\square$

THEOREM 11.3 ([**76**]). *Let* $(P, \tau)$ *be a real algebraic curve of type* $(g, k, 1)$. *Then two Arf functions on* $(P, \tau)$ *are topologically equivalent if and only if they have the same topological type* $(g, \tilde{\delta}, k_\alpha^\gamma)$.

Proof. Let $\omega_1$ and $\omega_2$ be two Arf functions on $(P, \tau)$, of type $(g, \tilde{\delta}, k_\alpha^\gamma)$. The ovals $P^\tau$ cut $P$ into two connected components $P_1$ and $P_2$. By assumption, the functions $\omega_1|_{P_1}$ and $\omega_2|_{P_1}$ have the same topological type and therefore, by Section 8 of Chapter 1, there is a homomorphism $\phi_1\colon P_1 \to P_1$, taking $\omega_1|_{P_1}$ to $\omega_2|_{P_1}$. The coincidence of the topological types of the functions $\omega_1$ and $\omega_2$ allows one to choose $\phi_1$ in such way that the ovals similar with respect to $\omega_1$ are taken to ovals similar with respect to $\omega_2$. Now let us set $\phi_2 = \tau\phi_1\tau\colon P_2 \to P_2$. Then $\phi_1 \cup \phi_2\colon P \to P$ commutes with $\tau$ and takes $\omega_1$ to $\omega_2$.                                    $\square$

**3.** In the rest of this section, the words "spinor bundle" always refer to a spinor bundle of rank 1.

Theorem 5.1 establishes a one-to-one correspondence between the spinor bundles on a real algebraic curve $(P, \tau)$ and the Arf functions on it. The *type of a spinor bundle* is the type of the corresponding Arf function.

The *moduli space* of spinor bundles on real algebraic curves is the space of pairs $((P, \tau), (e, E))$, where $(P, \tau)$ is a real algebraic curve and $(e, E)$ is a spinor bundle on $(P, \tau)$. By Theorem 5.1, there are only finitely many spinor bundles on a Riemann surface and therefore the topology of the moduli space of real curves induces a topology in the moduli space of spinor bundles on them.

THEOREM 11.4 ([**67**]). *The space of spinor bundles on nonseparating real algebraic curves decomposes into connected components* $S_p(g, \delta, k_\alpha)$, *where* $(g, \delta, k_\alpha)$ *is an arbitrary topological type of a nonspecial Arf function on a nonseparating real curve. Each of the connected components* $S(g, \delta, k_\alpha)$ *is diffeomorphic to*

$$\mathbb{R}^{3g-3} / \operatorname{Mod}_{g,\delta,k_\alpha}$$

*(where* $\operatorname{Mod}_{g,\delta,k_\alpha}$ *is a discrete group of automorphisms) and is a* $\left(\binom{k_0}{k} \cdot 2^{g-1}\right)$-*sheeted covering of* $M_{g,k,0}$, *where* $k = k_0 + k_1$.

Proof. By definition, a sequential set

$$V = (C_1, \ldots, C_{g+1}) \in \operatorname{Aut}(\Lambda)$$

of type $(0, g + 1)$ is associated to each $\psi \in \tilde{T}_{0,g+1}$, which, together with the elements

$$\hat{C}_i = \begin{cases} \overline{C}_i & \text{for } i \leqslant k, \\ \tilde{C}_i & \text{for } i > k, \end{cases}$$

generates a real Fuchsian group $\tilde{\Gamma} = \Gamma_{\psi,g+1}^k$. On a real curve $(P, \tau) = [\tilde{\Gamma}]$ consider the basis of homology $\{c_i, d_i \ (i = 1, \ldots, g)\} \in H_1(P, \mathbb{Z}_2)$ corresponding to the shifts $\{C_i, D_i = \tilde{C}_{g+1}\hat{C}_i \ (i = 1, \ldots, g)\}$. Consider the nonspecial real Arf function $\omega = \omega_\psi$ defined by the following conditions: $\omega(c_i) = 0$ for $i \leqslant k_0$, $\omega(c_i) = 1$ for $i > k_0$, $\omega(d_i) = 0$ for $i < g$, $\omega(d_g) = \delta$. By Theorem 3.2, we have $k_0 \equiv g + 1 \pmod{2}$, which immediately implies that $\omega$ is a nonspecial Arf function of type $(g, \delta, k_\alpha)$. By Theorem 5.1, associated to this function is a spinor bundle $\Omega(\psi) \in S_p(g, \delta, k_\alpha)$. The correspondence $\psi \mapsto \Omega(\psi)$ induces a mapping $\Omega \colon T_{0,g+1} \to S_p(g, \delta, k_\alpha)$.

Let us prove that $\Omega(T_{0,g+1}) = S_p(g, \delta, k_\alpha)$. Indeed, according to Theorem 2.2 the mapping

$$\Psi \colon T_{0,g+1} \xrightarrow{\ \Omega\ } S_p(g, \delta, k_\alpha) \xrightarrow{\ \Phi\ } M_{g,k,0},$$

where $\Phi$ is the natural projection, is such that

$$\Psi(T_{0,g+1}) = M_{g,k,0}.$$

A fiber of $\Psi$ is represented by the group $\mathrm{Mod}_{g,k,0}$ of all autohomeomorphisms of the curve $(P, \tau)$, i.e., of those autohomeomorphisms of $P$ that commute with $\tau$. By Theorem 11.2, this group $\mathrm{Mod}_{g,k,0}$ acts transitively on the set of nonsingular real Arf functions of type $(g, \delta, k_\alpha)$, and therefore, by Theorem 5.1, it acts transitively on the fibers $\Phi^{-1}((P, \tau))$. Hence,

$$\Omega(T_{0,g+1}) = S_p(g, \delta, k) \quad \text{and} \quad S_p(g, \delta, k_\alpha) = T_{0,g+1}/\mathrm{Mod}_{g,\delta,k_\alpha},$$

where

$$\mathrm{Mod}_{g,\delta,k_\alpha} \subset \mathrm{Mod}_{g,k,0}.$$

According to Section 4 of Chapter 1, the space $T_{0,g+1}$ is diffeomorphic to $\mathbb{R}^{3g-3}$. By Theorem 3.2, the index of the subgroup $\mathrm{Mod}_{g,\delta,k_\alpha}$ in $\mathrm{Mod}_{g,k_0+k_1,0}$ is $\binom{k_0}{k}2^{g-1}$.                                         $\square$

THEOREM 11.5 ([**67**]). *The space of spinor bundles on separating real algebraic curves decomposes into connected components $S_p(g, \tilde{\delta}, k_\alpha^\gamma)$, where $(g, \tilde{\delta}, k_\alpha^\gamma)$ is an arbitrary topological type of Arf functions on a real separating algebraic curve. Each connected component $S_p(g, \tilde{\delta}, k_\alpha^\gamma)$ is homeomorphic to $\mathbb{R}^{3g-3}/\mathrm{Mod}_{g,\tilde{\delta},k_\alpha^\gamma}$ (where $\mathrm{Mod}_{g,\tilde{\delta},k_\alpha^\gamma}$ is a discrete group of diffeomorphisms) and covers $M_{g,k,1}$ with the number of sheets equal to $\binom{k}{k_0} \cdot \binom{k_0}{k_0^0} \cdot \binom{k_1}{k_1^0} \cdot 2^{\tilde{g}-2} \times (2^{\tilde{g}} + m)$, where $m = 2^{\tilde{g}}$ if $k_1 > 0$, $m = 1$ if $\tilde{\delta} = 0$, $m = -1$ if $k_1 = 0$, and $\tilde{\delta} = 1$, $k_\alpha = k_\alpha^0 + k_\alpha^1$, $k = k_0 + k_1$, $g = 2\tilde{g} + k - 1$.*

Proof. By definition, associated to each $\psi \in \tilde{T}_{\tilde{g},k}$ is a sequential set $V = \{A_i, B_i \ (i = 1, \ldots, \tilde{g}), \ C_i \ (i = 1, \ldots, k)\}$ of type $(\tilde{g}, k)$, which, together with $\overline{C}_i \ (i = 1, \ldots, k)$, generates the real Fuchsian group $\tilde{\Gamma} = \Gamma_\psi^k$. On the real algebraic curve $(P, \tau) = [\tilde{\Gamma}]$ consider the homology basis $\{a_i, b_i, a_i', b_i' \ (i = 1, \ldots, \tilde{g}), c_i, d_i \ (i = 1, \ldots, k-1)\} \in H_1(P, \mathbb{Z}_2)$ generated by the shifts

$$\{A_i, B_i, \overline{C}_k A_i \overline{C}_k, \overline{C}_k B_i \overline{C}_k \ (i = 1, \ldots, \tilde{g}), C_i, \overline{C}_k \overline{C}_i \ (i = 1, \ldots, k-1)\}.$$

For definiteness, let $k_1^1 > 0$ (the other cases are treated similarly). Consider the real Arf function $\omega = \omega_\psi$ defined by the following conditions: (1) $\omega(a_i) = \omega(b_i) = \omega(a_i') = \omega(b_i') = \varepsilon_i$, where $\varepsilon_i = 0$ for $i < \tilde{g}$ and $\varepsilon_i = \tilde{\delta}$ for $i = \tilde{g}$; (2) $\omega(c_i) = 0$ for $i \leqslant k_0$, $\omega(c_i) = 1$ for $i > k_0$; (3) $\omega(d_i) = 0$ for $i = k_0^0 + 1, \ldots, k_0$ or $i = k_0 + k_1^0 + 1, \ldots, k - 1$ and $\omega(d_i) = 1$ otherwise. By Theorem 5.1, a spinor bundle $\Omega(\psi) \in S_p(g, \tilde{\delta}, k_\alpha^\gamma)$ is associated to this function. The rest of the proof coincides with the corresponding part of the proof of Theorem 11.4.
$\square$

## 12. Real algebraic $N = 1$ supercurves and their moduli space

We recall some definitions (cf. Section 11 of Chapter 1).

Let $L = L(\mathbb{K})$ be the Grassmann algebra over a field $\mathbb{K}$, with infinitely many generators $1, \ell_1, \ell_2, \ldots$. An element $a \in L(\mathbb{K})$ is a finite linear combination of monomials $\ell_{i_1} \wedge \cdots \wedge \ell_{i_n}$ with coefficients in $\mathbb{K}$,

$$a = a^\sharp + \sum a_i \ell_i + \sum_{ij} a_{ij} \ell_i \wedge \ell_j + \cdots .$$

The correspondence $a \mapsto a^\sharp$ determines an epimorphism $\sharp \colon L(\mathbb{K}) \to \mathbb{K}$.

A monomial $\ell_{i_1} \wedge \cdots \wedge \ell_{i_n} \neq 0$ is said to be *even* if $n$ is even, and *odd* otherwise. Constants are even monomials. Linear combinations, with coefficients in $\mathbb{K}$, of even (respectively, odd) monomials form the vector space $L_0(\mathbb{K})$ of even (respectively, the vector space $L_1(\mathbb{K})$ of odd) elements of $L(\mathbb{K})$. The superanalog of a vector space is the set

$$\mathbb{K}^{(n|m)} = \{(z_1, \ldots, z_n | \theta_1, \ldots, \theta_m) \mid z_i \in L_0(\mathbb{K}), \theta_j \in L_1(\mathbb{K})\}.$$

For a field $\mathbb{K}$, we usually take either the field of complex numbers $\mathbb{C}$ or the field of real numbers $\mathbb{R}$.

The set

$$\Lambda^{NS} = \{(z | \theta_1, \ldots, \theta_N) \in \mathbb{C}^{(1|N)} \mid \operatorname{Im} z^\sharp > 0\}$$

is called the *upper $N$-super half-plane*. In this section we shall make use of 1-super half-planes $\Lambda^S = \Lambda^{1S}$. The automorphism group $\operatorname{Aut}(\Lambda^S)$ of a superdomain $\Lambda^S$ consists of transformations $A = A[a, b, c, d, \sigma | \varepsilon, \delta]$ of the form

$$A(z | \theta) = \left( \frac{az + b}{cz + d} - \frac{(ad - bc)(\varepsilon + \delta z)}{(cz + d)^2} \theta \,\middle|\, \frac{\sigma \sqrt{ad - bc}}{cz + d} \left( \theta + \varepsilon + \delta z + \frac{1}{2} \varepsilon \delta \theta \right) \right),$$

where $a, b, c, d \in L_0(\mathbb{R})$, $\sigma = \pm 1$, $\varepsilon, \delta \in L_1(\mathbb{R})$, $(ad - bc)^\sharp > 0$, and the notation $\sqrt{\Delta}$ is used for the element in $L_0(\mathbb{R})$ uniquely determined by the properties $(\sqrt{\Delta})^2 = \Delta$ and $(\sqrt{\Delta})^\sharp > 0$.

The correspondence

$$A \mapsto A^\sharp, \quad \text{where} \quad A^\sharp(z) = \frac{a^\sharp z + b^\sharp}{c^\sharp z + d^\sharp},$$

induces an epimorphism

$$\sharp \colon \operatorname{Aut}(\Lambda^S) \to \operatorname{Aut}(\Lambda).$$

The elements of the set $\mathrm{Aut}_2(\Lambda^S) = (\sharp)^{-1}(\mathrm{Aut}_2(\Lambda))$ are said to be *superhyperbolic*.

Associate to an automorphism $A = A[a, b, c, d, \sigma \,|\, \varepsilon, \delta]$ the matrix

$$\bar{J}(A) = \frac{\sigma}{\sqrt{a^\sharp d^\sharp - c^\sharp d^\sharp}} \begin{pmatrix} a^\sharp & b^\sharp \\ c^\sharp & d^\sharp \end{pmatrix} \in \mathrm{SL}(2, \mathbb{R}).$$

A subgroup $\Gamma \subset \mathrm{Aut}(\Lambda^S)$ is said to be *super-Fuchsian* if $\Gamma^\sharp = \sharp(\Gamma)$ is a Fuchsian group and $\sharp\colon \Gamma \to \Gamma^\sharp$ is an isomorphism. In this section we consider (without special mentioning) only super-Fuchsian groups consisting of superhyperbolic automorphisms $\Lambda^S$.

The quotient set $P = \Lambda^S/\Gamma$ is called an ($N = 1$) *super-Riemann surface* (or a Riemann supersurface). The correspondence $\bar{J}$ induces the pullback

$$J^*\colon \Gamma^\sharp \to \Gamma^* \subset \mathrm{SL}(2, \mathbb{R}).$$

The type of the corresponding Arf function $\omega_\Gamma = \omega_{\Gamma^*}$ on $P^\sharp = \Lambda/\Gamma^\sharp$ is called the *type of the supersurface*.

**1.** Now let $\widetilde{\mathrm{Aut}}(\Lambda^S)$ be the group generated by $\mathrm{Aut}(\Lambda^S)$ and the involutions

$$\sigma_\pm\colon (z\,|\,\theta) \mapsto (-\bar{z}\,|\, \pm\bar{\theta}).$$

If $C \in \mathrm{Aut}_2(\Lambda^S)$ is a hyperbolic automorphism, then there is an element $g \in \mathrm{Aut}(\Lambda^S)$ such that $g^{-1}Cg(z) = (\lambda z\,|\,\sqrt{\lambda}\theta)$, where $\lambda^\sharp > 0$. We set

$$\overline{C}^\pm = g\sigma_\pm g^{-1}, \quad \tilde{C}^\pm = \sqrt{C}\,\overline{C}_\pm \in \widetilde{\mathrm{Aut}}(\Lambda^S),$$

where $g^{-1}\sqrt{C}g(z\,|\,\theta) = (\sqrt{\lambda}z\,|\,\sqrt[4]{\lambda}\theta)$. Extend the correspondence $\sharp\colon \mathrm{Aut}(\Lambda^S) \to \mathrm{Aut}(\Lambda)$ to a correspondence $\sharp\colon \widetilde{\mathrm{Aut}}(\Lambda^S) \to \widetilde{\mathrm{Aut}}(\Lambda)$ by setting $\sharp(\sigma_\pm) = \sigma_\pm^\sharp\colon z \mapsto -\bar{z}$.

We call a subgroup $\tilde{\Gamma} \subset \widetilde{\mathrm{Aut}}(\Lambda^S)$ a *real super-Fuchsian group* if $\tilde{\Gamma}^\sharp$ is a real Fuchsian group. Associated to a real super-Fuchsian group $\tilde{\Gamma}$ are a super-Fuchsian group $\Gamma = \tilde{\Gamma} \cap \mathrm{Aut}(\Lambda^S)$, a Riemann supersurface $P = \Lambda^S/\Gamma$, and a *real algebraic supercurve* $[\Gamma] = (P, \tau)$, where $\tau = (\tilde{\Gamma} \setminus \Gamma)/\Gamma\colon P \to P$ is a superantiholomorphic involution. To a supercurve $(P, \tau)$, a real algebraic curve

$$\sharp(P, \tau) = (P^\sharp, \tau^\sharp) = [\tilde{\Gamma}^\sharp],$$

called the *underlying curve* of $(P, \tau)$, is associated. It is easy to see that $\omega_\Gamma$ is a real Arf function on $(P^\sharp, \tau^\sharp)$. Its topological type is called the *topological type* of the real supercurve $(P, \tau)$.

**2.** Let $t = (\tilde{g}, \delta, k_\alpha)$ be the topological type of a Riemann supersurface of genus $\tilde{g}$ with $k$ holes. Denote by $M^t$ the set of all such supersurfaces. It admits a "uniformization" by the space

$$T^t = \tilde{T}^t / \mathrm{Aut}(\Lambda^S),$$

where $\tilde{T}^t$ is the space of monomorphisms $\psi\colon \gamma_{\tilde{g},n} \to \mathrm{Aut}(\Lambda^S)$ (where $n = \tilde{g} + k$) such that $\psi(v_{\tilde{g},n})^\sharp$ is a sequential set of type $(g, k)$, $\Lambda^S/\psi(\gamma_{\tilde{g},n}) \in M^t$, and the group $\mathrm{Aut}(\Lambda^S)$ acts by conjugations.

A set $Q$ is said to be strongly diffeomorphic to $\mathbb{R}^{(p|q)}$ if there is an embedding $Q \subset \mathbb{R}^{(p|q)}$ such that $Q^\sharp$ is diffeomorphic to $\mathbb{R}^p$ and $Q = \sharp^{-1}(\sharp(Q))$. According to Section 12 of Chapter 1, $T^t$ is strongly diffeomorphic to

$$\mathbb{R}^{(p|q)}/\mathbb{Z}_2 = \mathbb{R}^{(6\tilde{g}+3k-6|4\tilde{g}+2k-4)}/\mathbb{Z}_2.$$

Here we have

$$M^t = (T^t)/\operatorname{Mod}_t,$$

where $\operatorname{Mod}_t$ is a discrete group.

THEOREM 12.1 ([**67**], [**73**]). *The moduli space of real algebraic supercurves with a nonseparating underlying curve decomposes into connected components $S(g, \delta, k_\alpha)$, where $(g, \delta, k_\alpha)$ is an arbitrary topological type of a nonspecial Arf function on a nonseparating real curve. Each of the connected components has the form*

$$S(g, \delta, k_\alpha) = T_{g,\delta,k_\alpha}/\operatorname{Mod}_{g,\delta,k_\alpha},$$

*where $T_{g,\delta,k_\alpha}$ is strongly diffeomorphic to $\mathbb{R}^{(3g-3|2g-2)}/\mathbb{Z}_2$ and $\operatorname{Mod}_{g,\delta,k_\alpha}$ is a discrete group.*

Proof. Set $t = (0, 0, k_0, g + 1 - k_0)$. By definition, to each $\psi \in \tilde{T}^t$ there corresponds a set

$$V = (C_1, \ldots, C_{g+1}) \in \operatorname{Aut}(\Lambda^S)$$

such that $V^\sharp = (C_1^\sharp, \ldots, C_{g+1}^\sharp)$ is a sequential set of type $(0, g + 1)$. Together with the elements

$$\hat{C}_i = \begin{cases} \overline{C}_i^+ & \text{for } i \leqslant k, \\ \tilde{C}_i^+ & \text{for } k < i < g, \\ \tilde{C}_g^+ & \text{for } i = g,\ \delta = 0, \\ \tilde{C}_i^- & \text{for } i = g,\ \delta = 1 \end{cases}$$

(where $k = k_0 + k_1$), this set generates a super-Fuchsian group $\tilde{\Gamma}$. On the real curve $(P^\sharp, \tau^\sharp) = [\tilde{\Gamma}^\sharp]$ consider the homology basis

$$\{c_i, d_i\ (i = 1, \ldots, g)\} \in H_1(P, \mathbb{Z}_2)$$

corresponding to the shifts

$$\{C_i, D_i = \tilde{C}_{g+1}\hat{C}_i\ (i = 1, \ldots, g)\}.$$

Then the Arf function $\omega = \omega_\Gamma$ satisfies the conditions $\omega(c_i) = 0$ for $i \leqslant k_0$, $\omega(c_i) = 1$ for $i > k_0$, $\omega(d_i) = 0$ for $i < g$ and $\omega(d_g) = \delta$. Therefore, the correspondence $\psi \mapsto [\tilde{\Gamma}]$ induces a mapping

$$\Omega\colon (\tilde{T}^t) \to S(g, \delta, k_\alpha).$$

Under this correspondence, mutually conjugate elements $\psi$ are taken to the same supercurve, and hence a mapping

$$\Omega\colon (T^t) \to S(g, \delta, k_\alpha)$$

is well defined.

Let us prove that $\Omega(T^t) = S(g, \delta, k_\alpha)$. Let

$$(P, \tau) \in S(g, \delta, k_\alpha).$$

Lemma 1.2 and Theorem 11.2 imply that there are simple closed contours $\{c_i, d_i \ (i = 1, \ldots, g)\}$ on $(P^\sharp, \tau^\sharp)$ such that (1) $\tau^\sharp(c_i) = c_i$, $(P^\sharp)^{\tau^\sharp} = \bigcup_{i=1}^k c_i$; (2) the elements of $H_1(P^\sharp, \mathbb{Z}_2)$ represented by these contours satisfy the conditions $\tau^*(d_i) = -d_i + c + \hat{c}_i$, where $c = \sum_{i=1}^g c_i$,

$$\hat{c}_i = \begin{cases} 0 & \text{for } i \leqslant k, \\ c_i & \text{for } i > k; \end{cases}$$

(3) the Arf function $\omega = \omega_\Gamma$ satisfies the conditions

$$\omega(c_i) = \begin{cases} 0 & \text{for } i \leqslant k_0, \\ 1 & \text{for } i > k_0, \end{cases}$$

$\omega(d_i) = 0$ for $i < g$, $\omega(d_g) = \delta$. The contours $\{c_i\}$ cut the surface $P^\sharp$ into connected components $P_1^\sharp$ and $P_2^\sharp$. Set $P_1 = \sharp^{-1}(P_1^\sharp)$. According to Section 12 of Chapter 1, we have $P_1 = \Lambda^s / \psi(\gamma_{0,g+1})$, where $\psi \in \tilde{T}^t$. Our constructions immediately imply that $\Omega(\psi) = (P, \tau)$ and $\Omega(\psi') = \Omega(\psi)$ if and only if $\psi' = \psi\alpha$, where $\alpha \in \mathrm{Mod}_{g,\delta,k_\alpha}$ and $\mathrm{Mod}_{g,\delta,k_\alpha}$ is the group from Theorem 11.5. $\quad\square$

THEOREM 12.2 ([67], [73]). *The moduli space of real algebraic super-curves with separating underlying curve is decomposed into connected components $S(g, \tilde{\delta}, k_\alpha^\gamma)$ corresponding to arbitrary topological types $t = (g, \tilde{\delta}, k_\alpha^\gamma)$ of Arf functions on separating real curves. Each of the connected components has the form*

$$T^t / \mathrm{Mod}_{g,\tilde{\delta},k_\alpha^\gamma},$$

*where $T^t$ is strongly diffeomorphic to $\mathbb{R}^{(3g-3|2g-2)}/\mathbb{Z}_2$ and $\mathrm{Mod}_{g,\tilde{\delta},k_\alpha^\gamma}$ is a discrete group.*

Proof. We set $k_0 = k_0^0 + k_0^1$, $k_1 = k_1^0 + k_1^1$, $k = k_0 + k_1$, $\tilde{g} = \frac{1}{2}(g + 1 - k)$, and $t = (\tilde{g}, \tilde{\delta}, k_\alpha)$. By definition, to each $\psi = \tilde{T}^t$ there corresponds a set $V = (A_i, B_i \ (i = 1, \ldots, \tilde{g}), C_i \ (i = 1, \ldots, k)) \subset \mathrm{Aut}(\Lambda^S)$ such that $V^\sharp = \{A_i^\sharp, B_i^\sharp \ (i = 1, \ldots, \tilde{g}), C_i^\sharp \ (i = 1, \ldots, k)\}$ is a sequential set of type $(\tilde{g}, k)$. Together with the elements

$$\hat{C}_i = \begin{cases} \overline{C}_j^+ & \text{for } i \leqslant k_0^0 \text{ or } k_0 < i \leqslant k_0 + k_1^0, \\ \overline{C}_i^- & \text{for } k_0^0 < i \leqslant k_0 \text{ or } i > k_0 + k_1^0, \end{cases}$$

the set $V$ generates a real super-Fuchsian group $\tilde{\Gamma}$. The correspondence $\psi \mapsto [\tilde{\Gamma}]$ determines a mapping $\Omega \colon T^t \to S(g, \tilde{\delta}, k_\alpha^\gamma)$. The rest of the proof repeats, with obvious modifications, the corresponding part of the proof of Theorem 12.1. $\quad\square$

## 13. Real algebraic $N = 2$ supercurves

**1.**    Recall some definitions from Section 13 of Chapter 1. Denote by $A[a, b, c, d, \ell \,|\, \varepsilon]$ the mapping $A \colon \Lambda^{2S} \to \Lambda^{2S}$ of the form

$$A(z \,|\, \theta_1, \theta_2) = \left( \frac{az + b + \delta^{11}\theta_1 + \delta^{12}\theta_2}{cz + d + \delta^{21}\theta_1 + \delta^{22}\theta_2} \,\right|$$
$$\left. \frac{\ell^{11}\theta_1 + \ell^{12}\theta_2 + \varepsilon^{11}z + \varepsilon^{12}}{cz + d + \delta^{21}\theta_1 + \delta^{22}\theta_2}, \frac{\ell^{21}\theta_1 + \ell^{22}\theta_2 + \varepsilon^{21}z + \varepsilon^{22}}{cz + d + \delta^{21}\theta_1 + \delta^{22}\theta_2} \right),$$

where $a, b, c, d \in L_0(\mathbb{R})$, $\ell \in \mathrm{GL}(2, L_0(\mathbb{R}))$, $\varepsilon^{ij}, \delta^{ij} \in L_1(\mathbb{R})$.

According to [**45**], the automorphism group $\mathrm{Aut}(\Lambda^{2S})$ of the superdomain $\Lambda^{2S}$ consists of transformations $A[a, b, c, d, \ell \,|\, \varepsilon]$ such that

$$\begin{pmatrix} -c & a \\ -d & b \end{pmatrix} \begin{pmatrix} \delta^{11} & \delta^{12} \\ \delta^{21} & \delta^{22} \end{pmatrix} = \begin{pmatrix} \varepsilon^{21} & \varepsilon^{11} \\ \varepsilon^{22} & \varepsilon^{12} \end{pmatrix} \begin{pmatrix} \ell^{11} & \ell^{12} \\ \ell^{21} & \ell^{22} \end{pmatrix}$$

and

$$ad - bc - \varepsilon^{11}\varepsilon^{12} - \varepsilon^{21}\varepsilon^{22} = \ell^{11}\ell^{22} + \ell^{21}\ell^{12} + \delta^{11}\delta^{22} + \delta^{12}\delta^{21} = \Delta,$$

where $\Delta^\sharp > 0$,

$$\ell^{11}\ell^{21} + \delta^{11}\delta^{21} = \ell^{12}\ell^{22} + \delta^{12}\delta^{22} = 0.$$

A direct calculation shows that an automorphism $A[a, b, c, d, \ell \,|\, \varepsilon]$ belongs to one of the following two types:

(1) (nontwisted) $(\ell^{12})^\sharp = (\ell^{21})^\sharp = 0$; $(\ell^{11}\ell^{22})^\sharp > 0$;

(2) (twisted) $(\ell^{11})^\sharp = (\ell^{22})^\sharp = 0$; $(\ell^{12}\ell^{21})^\sharp > 0$.

A nontwisted (respectively, twisted) automorphism is uniquely determined by the parameters $a$, $b$, $c$, $d$, $\varepsilon^{ij}$, $\ell^{11}$ (respectively, $a$, $b$, $c$, $d$, $E^{ij}$, $\ell^{12}$). These parameters can take arbitrary values such that $a, b, c, d, \ell^{ij} \in L_0(\mathbb{R})$, $\varepsilon^{ij} \in L_1(\mathbb{R})$, $(ad - bc)^\sharp > 0$, $(\ell^{11} + \ell^{12})^\sharp \ne 0$.

The correspondence $A \mapsto A^\sharp$, where

$$A = A[a, b, c, d, \ell \,|\, \varepsilon], \quad A^\sharp(z) = \frac{a^\sharp z + b^\sharp}{c^\sharp z + d^\sharp},$$

induces an epimorphism $\sharp \colon \mathrm{Aut}(\Lambda^{2S}) \to \mathrm{Aut}(\Lambda)$. An element of the group $\mathrm{Aut}_2(\Lambda^{2S}) = (\sharp)^{-1}(\mathrm{Aut}_2(\Lambda))$ is said to be *superhyperbolic*.

A subgroup $\Gamma \subset \mathrm{Aut}(\Lambda^{2S})$ is said to be $N = 2$ *super-Fuchsian* if $\Gamma^\sharp = \sharp(\Gamma)$ is a Fuchsian group and $\sharp \colon \Gamma \to \Gamma^\sharp$ is an automorphism. In this section we consider (without special stipulation) only $N = 2$ super-Fuchsian groups consisting of superhyperbolic automorphisms $\Lambda^{2S}$.

Associate to an automorphism $A = A[a, b, c, d, \ell \,|\, \varepsilon]$ the matrix

$$\bar{J}(A) = \frac{\sigma}{\sqrt{a^\sharp d^\sharp - b^\sharp c^\sharp}} \begin{pmatrix} a^\sharp & b^\sharp \\ c^\sharp & d^\sharp \end{pmatrix} \in \mathrm{SL}(2, \mathbb{R}),$$

where $\sigma = \sigma(A) = \mathrm{sign}(\ell^{11} + \ell^{12} + \ell^{21} + \ell^{22})^\sharp$.

If $\Gamma \in \mathrm{Aut}(\Lambda^{2S})$ is an $N = 2$ super-Fuchsian group, then the correspondence $\bar{J} \colon \Gamma \to \mathrm{SL}(2, \mathbb{R})$ is a monomorphism, whence defining the pullback $J^* \colon \Gamma^\sharp \to \bar{J}(\Gamma)$.

Let $\Gamma \subset \mathrm{Aut}(\Lambda^{2S})$ be an $N = 2$ super-Fuchsian group. The quotient set $\Lambda^{2S}/\Gamma$ is called an $N = 2$ *super-Riemann surface*, or an $N = 2$ *Riemann supersurface*. Two $N = 2$ supersurfaces $P_1 = \Lambda^{2S}/\Gamma_1$ and $P_2 = \Lambda^{2S}/\Gamma_2$ are considered to be the same if the subgroups $\Gamma_1$ and $\Gamma_2$ are conjugate in $\mathrm{Aut}(\Lambda^{2S})$. The projections $\sharp \colon \Lambda^{2S} \to \Lambda$ and $\sharp \colon \Gamma \to \Gamma^\sharp$ determine the projection $\sharp \colon P \to P^\sharp = \Lambda/\Gamma^\sharp$.

According to Section 7 of Chapter 1, the pullback $J^*$ determines an Arf function

$$\omega_P^1 \colon H_1(P^\sharp, \mathbb{Z}_2) \to \mathbb{Z}_2.$$

Define two functions $\Omega_i = \Omega_i(\Gamma) \colon \Gamma \to \mathbb{Z}_2 = \{0, 1\}$ $(i = 1, 2)$ by setting

$$\Omega_1(A) = \begin{cases} 0 & \text{if } \sum_{i,j\in\{1,2\}}(h^{ij})^\sharp < 0, \\ 1 & \text{otherwise}, \end{cases}$$

and

$$\Omega(A) = \Omega_1(A) + \Omega_2(A) = \begin{cases} 0 & \text{if } h^{12} = h^{21} = 0, \\ 1 & \text{if } h^{11} = h^{22} = 0. \end{cases}$$

It is easy to see that $\Omega_1$ induces $\omega_P^1$ and $\Omega$ is a homomorphism inducing a homomorphism $\omega_P^0 \colon H_1(P^\sharp, \mathbb{Z}_2) \to \mathbb{Z}_2$. Denote by $\omega_P^2$ the Arf function $\omega_P^1 + \omega_P^0$ induced by $\Omega_2$.

An $N = 2$ super-Riemann surface $P$ is said to be *nontwisted* if $\omega_P^0 = 0$. The *topological type* of such a surface is the topological type $(g, \delta, k_\alpha)$ of the Arf function $\omega_P^1 = \omega_P^2$. If $\omega_P^0 \neq 0$, then the Riemann surface is said to be *twisted*. The *topological type* of such a surface is the topological type $(g, \delta_1, \delta_2, k_{\alpha\beta})$ of the pair of Arf functions $(\omega_P^1, \omega_P^2)$, where $\delta_i = \delta(P^\sharp, \omega_i)$ and $k_{\alpha\beta}$ is the number of holes $c_i$ of the surface $P^\sharp$ such that $\omega_1(c_i) = \alpha$, $\omega_2(c_i) = \beta$ (cf. Section 8 of Chapter 1).

**2.**     The $N = 2$ superanalog of the group $\widetilde{\mathrm{Aut}}(\Lambda^\sharp)$ is the group $\widetilde{\mathrm{Aut}}(\Lambda^{2S})$ generated by $\mathrm{Aut}(\Lambda^{2S})$ and the transformation $\sigma \colon (z|\theta_1, \theta_2) \mapsto (-\bar{z}|\bar{\theta}_1, \bar{\theta}_2)$. Extend $\sharp \colon \mathrm{Aut}(\Lambda^{2S}) \to \mathrm{Aut}(\Lambda)$ to a homomorphism $\sharp \colon \widetilde{\mathrm{Aut}}(\Lambda^{2S}) \to \widetilde{\mathrm{Aut}}(\Lambda)$ by setting $\sharp(\sigma) \colon z \mapsto -\bar{z}$. A subgroup $\tilde{\Gamma} \subset \widetilde{\mathrm{Aut}}(\Lambda^{2S})$ is called a *real $N = 2$ super-Fuchsian group* if $\Gamma = \tilde{\Gamma} \cap \mathrm{Aut}(\Lambda^{2S})$ is an $N = 2$ super-Fuchsian group, $\tilde{\Gamma} \neq \Gamma$, and $\Lambda^\sharp/\Gamma^\sharp$ is a compact surface. In this case the pair $(\Lambda^{2S}/\Gamma, \tilde{\Gamma}/\Gamma)$ is called a *real algebraic $N = 2$ supercurve*.

Real $N = 2$ supercurves $(\Lambda^{2S}/\Gamma_1, \tilde{\Gamma}_1/\Gamma_1)$ and $(\Lambda^{2S}/\Gamma_2, \tilde{\Gamma}_2/\Gamma_2)$ are considered to be the same if there is $h \in \widetilde{\mathrm{Aut}}(\Lambda^{2S})$ such that $\tilde{\Gamma}_2 = h\tilde{\Gamma}_1 h^{-1}$. The projection $\sharp$ takes a real supercurve $(P, \tau) = (\Lambda/\Gamma, \tilde{\Gamma}/\Gamma)$ to the real curve $(P^\sharp, \tau^\sharp) = (\Lambda^\sharp/\Gamma^\sharp, \tilde{\Gamma}^\sharp/\Gamma^\sharp)$.

Let $(P, \tau) = (\Lambda^{2S}/\Gamma, \tilde{\Gamma}/\Gamma)$ be a real $N = 2$ supercurve, and let $C \subset \Gamma$ correspond to an oval or to an invariant contour (not intersecting an oval) $c$. Replacing $\Gamma$, if necessary, by a conjugate group, we may assume that $C(z|\theta_1, \theta_2) = (\lambda z|h^1\theta_j, h^2\theta_{3-j})$. In this case the group $\tilde{\Gamma}$ contains an element $S_C$ of the form $S_C(z|\theta_1, \theta_2) = (-\rho\bar{z}|l^1\bar{\theta}_i, l^2\bar{\theta}_{3-i})$, where $\rho^\sharp > 0$, $\ell^1\ell^2 = \rho^2$, and $\rho = 1$ if

$c$ is an oval, and $(S_C)^2 = C$ if $c$ is an invariant contour. We set $\mu(c) = 0$ if $i = 1$ and $\mu(c) = 1$ if $i = 2$.

If $\omega_1 = \omega_2$ (where $\omega_i = \omega_P^i$), then $\mu(c)$ is the same for all ovals and invariant contours (not intersecting ovals) $c$. This fact allows us to define the invariant $\mu(P, \tau) = \mu(c)$.

If $\omega_1 \neq \omega_2$, then the kernel of the homomorphism $\Omega \colon \Gamma \to \mathbb{Z}_2$ is a subgroup $\Gamma_*$ of index 2. On the surface $P_*^\sharp = \Lambda^\sharp / \Gamma_*^\sharp$, the involutions in the set $\{F = S_C \mid \mu(c) = \mu\}$ induce an involution $\tau_\mu^\sharp$ ($\mu \in \mathbb{Z}_2$). Put $\rho_\mu(P, \tau) = \epsilon(P_*^\sharp, \tau_\mu^\sharp)$.

The set $M(g, \epsilon)$ consists of real algebraic $N = 2$ supercurves $(P, \tau)$ such that $g(P^\sharp) = g$, $\epsilon(P^\sharp, \tau^\sharp) = \epsilon \in \mathbb{Z}_2$. The structure of an $N = 2$ supercurve determines two Arf functions $\omega_i = \omega_P^i \colon H_1(P^\sharp, \mathbb{Z}_2) \to \mathbb{Z}_2$. Let $\chi(P) = 0$ if $\omega_1 = \omega_2$ and $\chi(P) = 1$ if $\omega_1 \neq \omega_2$. The invariant $\chi \in \mathbb{Z}_2$ splits $M(g, \epsilon)$ into subsets $M(g, \epsilon, \chi) = \{(P, \tau) \in M(g, \epsilon) \mid \chi(P) = \chi\}$.

By Theorem 3.2, the number of ovals $c$ with the property $\omega_i(c) = 0$ has the same parity as $g + 1$. For $(P, \tau) \in M(g, 0, 0)$, denote by $k_\alpha(P, \tau)$ the number of ovals $c$ such that $\omega_1(c) = \omega_2(c) = \alpha \in \mathbb{Z}_2$. Split the set $M(g, 0, 0)$ into the subsets

$$M(g, 0, 0, k_\alpha, \delta, \mu) = \{(P, \tau) \in M(g, 0, 0) \mid$$
$$k_\alpha(P, \tau) = k_\alpha, \delta(\omega_1) = \delta(\omega_2) = \delta, \mu(P, \tau) = \mu\}.$$

For $(P, \tau) \in M(g, 0, 1)$, denote by $k_{\alpha\beta}^\mu(P, \tau)$ the number of ovals $c \subset P^\tau$ such that

$$\omega_1(c) = \alpha, \quad \omega_2(c) = \beta, \quad \mu(c) = \mu \in \mathbb{Z}_2.$$

Set

$$M(g, 0, 1, k_{\alpha\beta}^\mu, \delta_i, \rho_i) = \{(P, \tau) \in M(g, 0, 1) \mid$$
$$k_{\alpha\beta}^\mu(P, \tau) = k_{\alpha\beta}^\mu, \delta(\omega_i) = \delta_i, \rho_i(P, \tau) = \rho_i\}.$$

According to [66] and [12], $M(g, 0, 1, k_{\alpha\beta}^\mu, \delta_i, \rho_i) = 0$ if $\rho_1 = \rho_2 = 1$ or if $k_{01}^0 + k_{10}^0 + k_{01}^1 + k_{10}^1 > 0$ and $\rho_1 + \rho_2 > 0$.

Let $(P^\sharp, \tau^\sharp)$ be a real algebraic curve such that $\epsilon(P^\sharp, \tau^\sharp) = 1$ and let $\omega \colon H_1(P^\sharp, \mathbb{Z}_2) \to \mathbb{Z}_2$ be an Arf function such that $\omega(\tau^\sharp a) = \omega(a)$ for all $a \in H_1(P^\sharp, \mathbb{Z}_2)$. The ovals $c_1, \dots, c_k$ cut $P^\sharp$ into connected components $P_1^\sharp$ and $P_2^\sharp$. Set

$$\eta_\omega(P^\sharp, \tau^\sharp) = \delta(P_1^\sharp, \omega'),$$

where $\omega'$ is the restriction of $\omega$ to $P_1^\sharp$. In particular, $\eta_\omega(P^\sharp, \tau^\sharp) = 0$ if there is an oval $c$ such that $\omega(c) = 1$.

Set

$$M(g, 1, 0, k_\alpha^\gamma, \eta, \mu) = \{(P, \tau) \in M(g, 1, 0) \mid$$
$$k_\alpha^\gamma(P^\sharp, \tau^\sharp, \omega_1) = k_\alpha^\gamma, \eta_{\omega_1}(P, \tau) = \eta, \mu(P, \tau) = \mu\}.$$

Now let $(P, \tau) \in M(g, 1, 1)$. Denote by $k_{\alpha\beta}^{0\mu}(P, \tau)$ (respectively, $k_{\alpha\beta}^{1\mu}(P, \tau)$) the number of ovals $c_i$ similar to $c_1$ with respect to $\omega_1$ (respectively, not similar to $c_1$ with respect to $\omega_1$) such that $\omega_1(c_i) = \alpha$, $\omega_2(c_i) = \beta$, $\mu(c_i) = \mu$. The set of numbers $k_{\alpha\beta}^{\gamma\mu} = k_{\alpha\beta}^{\gamma\mu}(P, \tau)$ is well defined up to a permutation $k_{\alpha\beta}^{\gamma\mu} \mapsto k_{\alpha\beta}^{1-\gamma\mu}$ caused by the choice of $c_1$. Set

$$M(g, 1, 1, k_{\alpha\beta}^{\gamma\mu}, \eta_i) = \{(P, \tau) \in M(g, 1, 1) \mid k_{\alpha\beta}^{\gamma\mu}(P, \tau) = k_{\alpha\beta}^{\gamma\mu}, \eta_{\omega_i}(P, \tau) = \eta_i\}.$$

Hence we arrived at the following statement.

THEOREM 13.1 ([**82**]). (1) *The set* $M(g, \epsilon, 0)$ *of real algebraic supercurves* $(P, \tau)$ *of genus* $g$ *possessing the property* $\omega_1(P) = \omega_2(P)$ *decomposes into subsets* $M(g, 0, 0, k_\alpha, \delta, \mu)$, $M(g, 1, 0, k_\alpha^\gamma, \eta, \mu)$, *where* $\alpha, \gamma, \delta, \eta, \mu \in \mathbb{Z}_2$, $0 \leq k_0 + k_1 \leq g$, $1 \leq \sum_{\alpha\gamma} k_\alpha^\gamma \leq g + 1$, $\sum_{\alpha\gamma} k_\alpha^\gamma \equiv g + 1 \pmod 2$, $k_0 \equiv g + 1 \pmod 2$, $k_0^0 + k_0^1 \equiv g + 1 \pmod 2$ *and* $\eta = 0$ *for* $k_1^0 + k_1^1 > 0$. *Among these subsets, only* $M(g, 1, 0, k_\alpha^\gamma, \eta, \mu)$ *and* $M(g, 1, 0, k_\alpha^{1-\gamma}, \eta, \mu)$ *coincide.*

(2) *The set* $M(g, \epsilon, 1)$ *of real supercurves* $(P, \tau)$ *of genus* $g$ *possessing the property* $\omega_1(P) \neq \omega_2(P)$ *decomposes into subsets* $M(g, 0, 1, k_{\alpha\beta}^\mu, \delta_i, \rho_i)$, $M(g, 1, 1, k_{\alpha\beta}^{\gamma\mu}, \eta_i)$, *where* $\alpha, \beta, \gamma, \mu, i, \delta_i, \rho_i, \eta_i \in \mathbb{Z}_2$, $0 \leq \sum_{\alpha\beta\mu} k_{\alpha\beta}^{\gamma\mu} \leq g$, $1 \leq \sum_{\alpha\beta\gamma\mu} k_{\alpha\beta}^{\gamma\mu} \leq g + 1$, $\sum_{\alpha\beta\gamma\mu} k_{\alpha\beta}^{\gamma\mu} \equiv g + 1 \pmod 2$, $\sum_{\mu\beta} k_{0\beta}^\mu \equiv \sum_{\mu\alpha} k_{\alpha 0}^\mu \equiv \sum_{\gamma\mu\beta} k_{0\beta}^{\gamma\mu} \equiv \sum_{\gamma\mu\alpha} k_{\alpha 0}^{\gamma\mu} \equiv g + 1 \pmod 2$, $\rho_1 + \rho_2 < 2$, $\rho_1 = \rho_2 = 0$ *for* $k_{01}^0 + k_{01}^1 + k_{10}^0 + k_{10}^1 > 0$, $\eta_1 = 0$ *for* $\sum_{\beta\gamma\mu} k_{1\beta}^{\gamma\mu} > 0$ *and* $\eta_2 = 0$ *for* $\sum_{\alpha\gamma\mu} k_{\alpha 1}^{\gamma\mu} > 0$. *Among these subsets, only the sets* $M(g, 1, 1, k_{\alpha\beta}^{\gamma\mu}, \eta_i)$ *and* $M(g, 1, 1, k_{\alpha\beta}^{1-\gamma,\mu}, \eta_i)$ *coincide.*

## 14. Moduli spaces of real algebraic $N = 2$ supercurves

**1.**   Let $(P, \tau)$ be a real algebraic curve. A *double Arf function* on $(P, \tau)$ is a pair $(\omega, \alpha)$, where $\omega \colon H_1(P, \mathbb{Z}_2) \to \mathbb{Z}_2$ is an Arf function on $(P, \tau)$, and $\alpha \colon H_1(P, \mathbb{Z}_2) \to \mathbb{Z}_2$ is a homomorphism such that $\alpha\tau = \alpha$. Two double Arf functions $(\omega_1, \alpha_1)$ on $(P_1, \tau_1)$ and $(\omega_2, \alpha_2)$ on $(P_2, \tau_2)$ are said to be *topologically equivalent* if there is a homeomorphism $\phi \colon P_1 \to P_2$ such that $\phi\tau = \tau\phi$, $\omega_1 = \omega_2\phi$, $\alpha_1 = \alpha_2\phi$.

According to Section 13, a real algebraic $N = 2$ supercurve induces a double Arf function $(\omega_P, \alpha_P) = (\omega_P^1, \omega_P^0)$ on $(P^\sharp, \tau^\sharp)$.

THEOREM 14.1 ([**82**]). *Two real algebraic* $N = 2$ *supercurves* $P_1$ *and* $P_2$ *induce topologically equivalent double Arf functions if and only if their topological types* $P_1$ *and* $P_2$ *either coincide or differ from each other by the simultaneous change* $\mu \mapsto 1 - \mu$, $k_{\alpha\beta}^\mu \mapsto k_{\alpha\beta}^{1-\mu}$, *and* $k_{\alpha\beta}^{\gamma\mu} \mapsto k_{\alpha\beta}^{\gamma,1-\mu}$.

Proof. All topological invariants related to the supercurve $P$, except $\mu(c)$ for ovals and invariant contours $c$, are uniquely determined by a pair of Arf functions $(\omega_P^1, \omega_P^2)$ and are therefore preserved under the homeomorphisms $\phi \colon P_1^\sharp \to P_2^\sharp$ consistent with $\tau^\sharp$. Hence a topological equivalence of double Arf functions $(\omega_{P_i}^1, \omega_{P_i}^0)$ implies the restrictions on the types of $P_1$ and $P_2$ mentioned in the statement of the theorem.

The proof of the converse statement in the case $\varepsilon(P_i^\sharp, \tau_i^\sharp) = 1$ repeats that of Theorem 11.3. Now let us prove the converse statement for curves $(P_i^\sharp, \tau_i^\sharp)$ of type $(g, k, 0)$. Consider the standard bases existing according to Theorem 1.1,

$$\{a_j^i, b_j^i \ (i = 1, \ldots, 2r), c_j^i, d_j^i \ (i = 1, \ldots, s)\} \in H_1(P_j^\sharp, \mathbb{Z}_2),$$

where

$$(P_i^\sharp)^{\tau_i^\sharp} = \bigcup_{j=1}^{k} c_j^i, \ s - k \leqslant 1,$$

$$\tau_i^\sharp(a_j^i) = a_{2r+1-j}^i, \ \tau_i^\sharp(b_j^i) = -b_{2r+1-j}^i, \ \tau_i^\sharp(c_j^i) = c_j^i, \ \tau_i^\sharp(d_j^i) = -d_j^i + c^i,$$

where $c^i = \sum_{j=1}^{s} c_j^i$ and

$$\tau_i^\sharp(d_s^i) = -d_s^i + c^i + c_s^i \quad \text{for } s > k.$$

By Theorem 11.2, these bases can be chosen in such a way that

$$\omega_{P_1}^1(d_j^1) = \omega_{P_2}^1(d_j^2)$$

for all $j \leqslant s$. Under the assumptions of Theorem 14.1, one can choose a numbering of $c_j^i$ and $d_j^i$ such that

$$\omega_{P_1}^m(c_j^1) = \omega_{P_2}^m(c_j^2), \quad \omega_{P_1}^m(d_j^1) = \omega_{P_2}^m(d_j^2)$$

for all $j$ and $m = 1, 2$. Moreover, by Theorem 8.1 of Chapter 1, under the assumptions of Theorem 14.1 one can choose another basis of the same form preserving $c_j^i, d_j^i$, where

$$\omega_{P_1}^m(a_j^1) = \omega_{P_2}^m(a_j^2) \quad \text{and} \quad \omega_{P_1}^m(b_j^1) = \omega_{P_2}^m(b_j^2)$$

for all $j$. Then a homeomorphism $\phi\colon P_1^\sharp \to P_2^\sharp$ taking the first basis to the second induces a topological equivalence of the double Arf functions. $\qquad \square$

**2.** Recall the description of the moduli space of $N = 2$ super-Riemann surfaces (cf. Section 14 of Chapter 1).

Let $2t$ be the topological type of an $N = 2$ super-Riemann surface of genus $\tilde{g}$ with $k$ holes. Denote by $M$ the set of all such supersurfaces. It admits a "uniformization" by the space

$$T^{2t} = \tilde{T}^{2t} / \operatorname{Aut}(\Lambda^{2S}),$$

where $\tilde{T}^{2t}$ is the space of monomorphisms $\psi\colon \gamma_{\tilde{g},n} \to \operatorname{Aut}(\Lambda^{2S})$ (where $n = \tilde{g} + k$) such that $\psi(v_{\tilde{g},n})^\sharp$ is a sequential set of type $(g, k)$, $\Lambda^{2S} / \psi(\gamma_{\tilde{g},n}) \in M^{2t}$, and the group $\operatorname{Aut}(\Lambda^{2S})$ acts by conjugation (cf. Section 14 of Chapter 1). According to Section 14 of Chapter 1, $T^{2t}$ is strongly diffeomorphic to

$$\mathbb{R}^{(p|q)} / \mathbb{Z}_2 = \mathbb{R}^{(8\tilde{g}+4k-b(2t) \mid 8\tilde{g}+4k-8)} / (\mathbb{Z}_2)^2,$$

where $b(2t)$ is 8 for surfaces of twisted type and 7 otherwise. Here

$$M^{2t} = (T^{2t}) / \operatorname{Mod}_{2t},$$

where $\operatorname{Mod}_{2t}$ is a discrete group.

**3.** Now let us give a description of the moduli space of real $N = 2$ super-curves.

THEOREM 14.2 ([**82**]). (1) *The moduli space $M(g, \epsilon, 0)$ of real algebraic $N = 2$ supercurves $(P, \tau)$ of genus $g$ with the property $\omega_1(P) = \omega_2(P)$ decomposes into connected components $M(g, 0, 0, k_\alpha, \delta, \mu)$, $M(g, 1, k_\alpha^\gamma, \eta, \mu)$, where $\alpha, \gamma, \delta, \eta, \mu \in \mathbb{Z}_2$, $0 \le k = k_0 + k_1 \le g$, $1 \le \sum_{\alpha\gamma} k_\alpha^\gamma \le g + 1$, $k = \sum_{\alpha\gamma} k_\alpha^\gamma \equiv g + 1 \pmod 2$, $k_0 \equiv g + 1 \pmod 2$, $k_0^0 + k_0^1 \equiv g + 1 \pmod 2$ and $\eta = 0$ for $k_1^0 + k_1^1 > 0$. Among them, only the connected components $M(g, 1, 0, k_\alpha^\gamma, \eta, \mu)$ and $M(g, 1, 0, k_\alpha^{1-\gamma}, \eta, \mu)$ coincide. Each connected component $M(\chi)$ has the form $T(\chi)/\operatorname{Mod}(\chi)$, where $T(\chi)$ is strongly diffeomorphic to*

$$\mathbb{R}^{(4g-3+\mu k \mid 4g-4)}/(\mathbb{Z}_2)^2$$

*and $\operatorname{Mod}(\chi)$ is a discrete group.*

(2) *The moduli space $M(g, \epsilon, 1)$ of real algebraic $N = 2$ supercurves $(P, \tau)$ of genus $g$ with the property $\omega_1(P) \ne \omega_2(P)$ decomposes into connected components of the form $M(g, 0, 1, k_{\alpha\beta}^\mu, \delta_i, \rho_i)$, $M(g, 1, 1, k_{\alpha\beta}^{\gamma\mu}, \eta_i)$, where $\alpha, \beta, \gamma, \mu$, $i, \delta_i, \rho_i, \eta_i \in \mathbb{Z}_2$, $0 \le \sum_{\alpha\beta\mu} k_{\alpha\beta}^\mu \le g$, $1 \le \sum_{\alpha\beta\gamma\mu} k_{\alpha\beta}^{\gamma\mu} \le g + 1$, $\sum_{\alpha\beta\gamma\mu} k_{\alpha\beta}^{\gamma\mu} \equiv g + 1 \pmod 2$, $\sum_{\mu\beta} k_{0\beta}^\mu \equiv \sum_{\mu\alpha} k_{\alpha 0}^\mu \equiv \sum_{\gamma\mu\beta} k_{0\beta}^{\gamma\mu} \equiv \sum_{\gamma\mu\alpha} k_{\alpha 0}^{\gamma\mu} \equiv g + 1 \pmod 2$, $\rho_1 + \rho_2 < 2$, $\rho_1 = \rho_2 = 0$ for $k_{01}^0 + k_{01}^1 + k_{10}^0 + k_{10}^1 > 0$, $\eta_1 = 0$ for $\sum_{\beta\gamma\mu} k_{1\beta}^{\gamma\mu} > 0$ and $\eta_2 = 0$ for $\sum_{\alpha\gamma\mu} k_{\alpha 1}^{\gamma\mu} > 0$. Among these connected components, only the sets $M(g, 1, 1, k_{\alpha\beta}^{\gamma\mu}, \eta_i)$ and $M(g, 1, 1, k_{\alpha\beta}^{1-\gamma,\mu}, \eta_i)$ coincide. Each connected component $M(\chi)$ has the form $T(\chi)/\operatorname{Mod}(\chi)$, where $\operatorname{Mod}(\chi)$ is a discrete group and $T(\chi)$ is strongly diffeomorphic to $\mathbb{R}^{(4g-4+k^1 \mid 4g-4)}/(\mathbb{Z}_2)^2$ and $k^1$ equals either $\sum_{\alpha\beta\gamma} k_{\alpha\beta}^{\gamma 1}$ or $\sum_{\alpha\beta} k_{\alpha\beta}^1$.*

Proof. Let $\tilde{\Gamma}$ be a real $N = 2$ super-Fuchsian group, $(P, \tau) = [\tilde{\Gamma}]$, let $c$ be an oval or an invariant contour of $(P^\sharp, \tau^\sharp)$ not intersecting ovals, and let $C \subset \Gamma = \tilde{\Gamma} \cap \operatorname{Aut}(\Lambda^{2S})$ be the corresponding shift. Replacing $\tilde{\Gamma}$ by a conjugate group, if necessary, we may assume that $C(z \mid \theta_1, \theta_2) = (\rho z \mid \ell_1 \theta_i, \ell_2 \theta_{3-i})$. Denote by $\hat{C} \subset \tilde{\Gamma} \setminus \Gamma$ the element satisfying the following conditions: (1) $\hat{C} C \hat{C}^{-1} = C$; (2) $\hat{C}^2 = 1$, if $c$ is an oval; (3) $\hat{C}^2 = C$ if $c$ is an invariant contour. If $\mu(c) = 0$ and $c$ is an oval, then

$$\hat{C}(z \mid \theta_1, \theta_2) = (-\bar{z} \mid \pm \bar{\theta}_1, \pm \bar{\theta}_2).$$

If $\mu(c) = 0$ and $c$ is an invariant contour, then

$$\hat{C}(z \mid \theta_1, \theta_2) = (-\sqrt{\rho}\bar{z} \mid \pm \sqrt{|\ell_1|} \bar{\theta}_1, \pm \sqrt{|\ell_2|} \bar{\theta}_2).$$

If $\mu(c) = 1$ and $c$ is an oval, then

$$\hat{C}(z \mid \theta_1, \theta_2) = (-\bar{z} \mid h \bar{\theta}_2, h^{-1} \bar{\theta}_1).$$

If $\mu(c) = 1$ and $c$ is an invariant contour, then

$$\hat{C}(z \mid \theta_1, \theta_2) = (-\sqrt{\rho}\bar{z} \mid h \bar{\theta}_2, \sqrt{\rho} h^{-1} \bar{\theta}_1).$$

The rest of the proof repeats that of Theorems 12.1 and 12.2 with the replacement of the spaces $T^t$ by $T^{2t}$ and Theorems 11.2, 11.3 by Theorem 14.1. The only essential difference arises when we assign to a shift $C_i$ in the set

$$\{A_i, B_i \ (i=1,\ldots,\tilde{g}), C_i \ (i=1,\ldots,m)\} = \psi(v_{\tilde{g},m}), \quad \psi \in T^{2t}$$

the mapping $\hat{C}_i$. The above argument shows that if $\mu(c_i) = 0$ for the contour $c_i$ corresponding to $C_i$, then the shift $C_i$ determines the mapping $\hat{C}_i$ up to the same ambiguity as in the case $N=1$ (Section 12). For $\mu(c_i) = 1$ the choice of $\hat{C}_i$ depends on a single additional arbitrary parameter $h \in L_0(\mathbb{R})$. However, if $c_i$ is not an oval, then the condition $\hat{C}_i^2 = C_i$ picks one of the parameters belonging to $L_0(\mathbb{R})$, on which an arbitrary $C_i \in \mathrm{Aut}_2(\Lambda^{2S})$ depends. This property determines the dimensions of the supervector spaces uniformizing the connected components of the moduli spaces of real $N=2$ supercurves.

$\square$

# CHAPTER 3

# Spaces of Meromorphic Functions
# on Complex and Real Algebraic Curves

## 1. Coverings with simple critical points

**1.** Let $f\colon P \to S$ be an $n$-sheeted ramified covering of the sphere $S$ by a compact surface $P$ of genus $g$, let $A_f \subset P$ be the set of its critical (ramification) points and let $B_f = f(A_f)$ be the set of its critical values.

A critical value $s \in \mathbb{R}$ is said to be *simple* if its preimage $f^{-1}(s)$ consists of $(n-1)$ points, and it is said to be *degenerate* otherwise. A critical point $p \in A_f$ is said to be *simple* (respectively, *degenerate*) if its image $f(p)$ is a simple (respectively, degenerate) critical value.

A *segment* is a subset on a surface, homeomorphic to a straight segment, and a *contour* is just a simple closed contour. A *homotopy*, or a *deformation*, of a segment is a homotopy with fixed ends.

A closed contour $a \subset P$ is said to be *vanishing* with respect to $f$ if $f(a)$ is a segment intersecting the set $B_f$ exactly at the ends, and such that one of the ends is a simple critical value.

A subset $\ell \subset P$ is said to be *one-sheeted* if it is connected and the restriction $f|_\ell$ is one-to-one.

A *star centered at a point* $p^* \subset P - f^{-1}(B_f)$ is a set of one-sheeted segments $L = \{\ell_1, \ldots, \ell_r\}$, where the segment $\ell_i$ has the ends at $p^*$ and $p_i \in A_f$, such that (1) $f(\ell_i) \cap B_f = f(p_i)$ $(i = 1, \ldots, r)$; (2) $f(\ell_i) \cap f(\ell_j) = f(p^*)$ $(i \neq j = 1, \ldots, r)$; (3) for $g > 0$ all $p_i$ are simple critical points; (4) for $g = 0$ there is at most one degenerate critical point among the points $p_i$.

LEMMA 1.1. *Let $s_0 \in B_f$ and suppose all critical values in the set $B_f - \{s_0\}$ are simple. Then there is a vanishing contour in $P$. If, moreover, $g > 0$, then there is a vanishing contour not containing points in the set $f^{-1}(s_0)$.*

Proof. Let $p^* \subset P - f^{-1}(B_f)$. Choose among all stars centered at $p^*$ a star $L = \{\ell_1, \ldots, \ell_r\}$ that is not a part of any other star. For each $i$ the set $f^{-1}(f(\ell_i))$ contains a connected component $\bar{\ell}_i$ that contains the point $p^*$ and the critical point $p_i \in A_f$. Besides the point $p^*$, the connected component $\bar{\ell}_i$ contains at least one more point $p_i^*$ of $f^{-1}(f(p^*))$.

Suppose that all the points $p_i^*$ are pairwise distinct. Then $n - 1 \geqslant r$. On the other hand, by the Riemann–Hurwitz theorem, the number $v$ of simple critical points on $P - f^{-1}(s_0)$ satisfies the condition

$$v = 2(g + n - 1) - \sum_{p \in f^{-1}(s_0)} (v_p - 1),$$

where $v_p$ is the ramification index of $f$ at $p$. Therefore,

$$v \geqslant 2(g - 1) + 2n - n + 1 \geqslant 2g + r.$$

Hence there is a critical point $p \neq p_j$ $(i = 1, \ldots, r)$ such that the set $(p_1, \ldots, p_r, p)$ contains at most one degenerate critical point and $p \notin f^{-1}(s_0)$ for $g > 0$.

Since $p_i^* \neq p_j^*$ for $i \neq j$, the set $f^{-1}(f(L))$ can be continuously contracted along itself to a subset of $f^{-1}(f(p^*))$. Hence the set $P - f^{-1}(f(L))$ is connected. Therefore, there is a one-sheeted segment $\ell \in P$ with ends at $p^*$ and $p$ such that

$$f(\ell) \cap (B_f \cup L) = f(p) \cup f(p^*).$$

This means that the star $L$ is not maximal. This proves that there are coinciding points among $p_1^*, \ldots, p_k^*$, say, $p_1^* = p_2^*$. But then the set $f^{-1}(\ell_1 \cup \ell_2)$ contains a vanishing contour $a$ with the critical points $p_1$ and $p_2$.  □

**2.**     Let $a \subset P$ be a vanishing contour. We add to the set $P - a$ two copies of the contour $a$. As a result, we obtain a surface with two boundary components $a_1$ and $a_2$, which is either connected or consists of two connected components. Let us identify the points of $a_i$ taken by $f$ to the same point $(i = 1, 2)$. As a result we obtain either one or two surfaces without boundary. Let $\overline{P}$ be one of these surfaces. The covering $f$ induces a covering $\bar{f} : \overline{P} \to S$. We call the surface $\overline{P}$ together with the mapping $\bar{f}$ the *reduction of the surface $P$ and the mapping $f$ modulo the contour $a$*. Corresponding to the contours $a_i$ $(i = 1, 2)$ is one or two one-sheeted (with respect to $\bar{f}$) segments $(l_a^i)$ $(i = 1, 2)$ on the surface $\overline{P}$, which we call the *trace of the reduction*. The segment $s_a = \bar{f}(\ell_a^i) = f(a)$ is called the *reduction segment*.

LEMMA 1.2. *Let $g > 0$, $s_0 \in B_f$ and suppose all critical values in the set $B_f - \{s_0\}$ are simple. Then there is a vanishing contour $a$ such that $s_0 \in f(a)$ and $P - a$ is a connected set.*

Proof. The proof proceeds by induction on the number $n$ of sheets. For $n = 2$ all critical points of the covering are simple. The preimage $a = f^{-1}(\ell)$ of any segment $\ell \in S - s_0$ connecting critical values and containing no other critical values is a vanishing contour and the surface $P - a$ is connected.

Suppose the assertion of Lemma 1.2 is proved for $n < d$ and $n = d > 2$. By Lemma 1.1, there is a vanishing contour $a \subset P$ such that $s_0 \notin f(a)$. Suppose that the set $P - a$ is disconnected. Then there is a reduction $\bar{f} : \overline{P} \to S$ of $f$ modulo the contour $a$ such that the genus $\bar{g}$ of the surface $\overline{P}$ is positive. Applying the induction hypothesis we find a contour $\bar{b} \subset \overline{P}$ vanishing with respect to $\bar{f}$ and such that $\overline{P} - \bar{b}$ is a connected set. Since $s_a \cap B_{\bar{f}} = \varnothing$, there is a homotopy in the set $S - B_{\bar{f}}$ of the segment $\bar{f}(\bar{b})$ into a segment $\ell$ such that $\ell \cap s_a = \varnothing$. The preimage $\bar{f}^{-1}(\ell)$ contains a contour $b$ homotopic to $\bar{b}$.

It is easy to see that $b$ is a closed contour of the surface $P$ and $P - b$ is a smooth set. □

**3.** A segment $\ell \subset S$ is said to be *vanishing* with respect to $f$ if the preimage $f^{-1}(\ell)$ contains a vanishing closed contour $a_\ell$. For a covering $f$ having at most one degenerate critical value, the contour $a_\ell$ is defined uniquely and it is called the *contour of the vanishing segment* $\ell$.

We say that a vanishing segment $\ell$ is *cutting* if the set $P - a_\ell$ consists of two connected components, one of which, denoted $U_\ell$ and called the *film of the segment* $\ell$, is one-sheeted. The reduction $\bar{f} \colon \overline{P} \to S$ modulo the contour $a_\ell$, where $\overline{P} \supset P - U_\ell$, is called the *reduction modulo the cutting segment* $\ell$. The trace $b_\ell$ of this reduction is called the *trace of the segment* $\ell$.

The reduction modulo a set of disjoint cutting segments and its trace are defined similarly.

LEMMA 1.3. *Suppose a covering $f$ has at most one degenerate critical value, $g = 0$ and $n > 1$. Then a cutting segment exists.*

Proof. By Lemma 1.1, there is at least one vanishing contour on the surface $P$. Choose a vanishing contour $a$ such that one of the reductions $\bar{f} \colon \overline{P} \to S$ modulo $a$ has the minimal number of sheets $\bar{n}$. Suppose that $\bar{n} > 1$. Then, by Lemma 1.1, there is a contour $\bar{b}$ vanishing with respect to $\bar{f}$. At least one of the ends of the reduction segment $s_a$ does not belong to $B_{\bar{f}}$. Therefore, there is a monodromy along the set $S - B_{\bar{f}}$ taking the segment $f(\bar{b})$ to a segment $\ell$ having no common internal points with $s_a$. The preimage $\bar{f}^{-1}(\ell)$ of this segment contains a contour $b$ vanishing with respect to $\bar{f}$ and cutting the surface $\overline{P}$ into two parts. The trace of the reduction $\ell_a$ is totally contained in a part $U$. Therefore, $b$ is a closed contour on the surface $P$ cutting off a domain $U' = \overline{P} - U$. The number of sheets of the restriction $f|_{U'} = \bar{f}|_{U'}$ is smaller than $\bar{n}$. Therefore, one of the reductions of $f$ modulo the vanishing contour $b$ has less than $\bar{n}$ sheets. The contradiction thus obtained proves that $\bar{n} = 1$ and $s_a$ is a cutting segment. □

LEMMA 1.4. *Let $g = 0$, $s_0 \in B_f$ and suppose all critical values in the set $B_f - s_0$ are simple. Then there is a cutting segment $\ell \ni s_0$.*

Proof. The proof proceeds by induction on $n$. For $n = 2$ the assertion of Lemma 1.4 is obvious. Suppose the assertion is proved for $n < d$ and $n = d \geqslant 3$. Then, by Lemma 1.3, there is a cutting segment $\ell_1$. Let $s_0 \notin \ell_1$. Consider the reduction $\bar{f} \colon \overline{P} \to S$ of $f$ modulo $\ell_1$. Let $a$ be the contour of the segment $\ell_1$ and let $\ell_a \subset \overline{P}$ be the trace of the reduction. By the induction hypothesis, there is a segment $\ell' \ni s_0$ cutting with respect to the mapping $\bar{f}$. Since $\ell_1 \cap B_{\bar{f}} = \varnothing$, there is a monodromy along the set $S - B_{\bar{f}}$ taking the segment $\ell'$ to the segment $\ell$ not intersecting $\ell_1$ and such that the film $\overline{U}_\ell \subset \overline{P}$ of the segment $\ell$ with respect to the mapping $\bar{f}$ does not contain the segment $\ell_a$. But then $\ell$ is a cutting segment with respect to $f$. □

LEMMA 1.5. *Let $n > 2$, $s_0 \in B_f$ and suppose all critical values in the set $B_f - s_0$ are simple. Then there is a cutting segment $\ell \ni s_0$.*

Proof. The proof proceeds by induction on $g$. For $g = 0$ the assertion of Lemma 1.5 follows from Lemma 1.4. Suppose Lemma 1.5 is proved for $g < d$ and $g = d > 0$. Then, by Lemma 1.2, there is a vanishing contour $a \subset P$ not containing degenerate critical values and such that the set $P - a$ is connected. Let $\bar{f} \colon \overline{P} \to S$ be the reduction of the covering $f$ modulo the contour $a$. By the induction hypothesis, there is a vanishing segment $\bar{\ell} \ni s_0$ with respect to $\bar{f}$ on the sphere $S$.

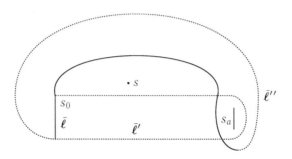

FIGURE 3.1.1

Since $n > 2$, the set $B_f$ contains at least three points. Let $U \subset S$ be the connected domain containing $s_a$, $\bar{\ell}$ and $s \in B_f - \ell$, and $U \cap B_{\bar{f}} \subset \ell \cup s$. Consider segments $\bar{\ell}'$ and $\bar{\ell}'' \subset U$ situated as shown in Fig. 3.1.1. The segments $\bar{\ell}$, $\bar{\ell}'$ and $\bar{\ell}''$ are vanishing segments with respect to the covering $\bar{f}$ and the films $U_{\bar{\ell}}$, $U_{\bar{\ell}'}$, $U_{\bar{\ell}''} \subset \overline{P}$ contain pairwise distinct preimages of the segment $s_a$. Therefore, at least one of the films of these segments does not contain the trace of the reduction of $\ell_a^i$ modulo the contour $a$. The segment corresponding to this film is a cutting segment with respect to the covering $f$. $\qquad\square$

**4.** The following assertion holds.

LEMMA 1.6. *Suppose all critical points of a mapping $f$ are simple, $s_1, \ldots, s_m \in S - B_f$ and $f^{-1}(s_k) = \{p_1^k, \ldots, p_n^k\}$ $(k = 1, \ldots, m)$. Then there is a set of vanishing segments $\ell_1, \ldots, \ell_{g+n-1}$ such that: (1) $\ell_i \cap \ell_j = \varnothing$ for $i \neq j$; (2) $l_1, \ldots, l_{n-2}$ are cutting segments; (3) the contours $a_i$ of the segments $\ell_{n+i-2}$ $(i = 1, \ldots, g + 1)$ together cut from the surface $P$ a connected component $U_{n-1} \subset P - \bigcup a_i$ that does not contain the films $U_i = U_{\ell_i}$ of the segments $\ell_i$ $(i = 1, \ldots, n - 2)$; (4) $p_i^k \subset U_i$ $(i = 1, \ldots, n - 1;\ k = 1, \ldots, m)$.*

Proof. We proceed by induction on $n$. For $n = 2$ the assertion of the lemma is obvious. Suppose the lemma is proved for $n < d$ and $n = d > 2$. Then, by Lemma 1.5, there is a cutting segment $\ell$. Let $\bar{f} \colon \overline{P} \to S$ be the reduction of $f$ modulo this segment, let $a$ be the contour of the segment, and let $\ell_a \subset \overline{P}$ denote the trace of the reduction. By the induction hypothesis, there are pairwise disjoint segments $\bar{\ell}_1,\ldots,\bar{\ell}_{g+n-2}$ vanishing with respect to $\bar{f}$ and such that $\bar{\ell}_1,\ldots,\bar{\ell}_{n-3}$ are cutting segments and the contours $\bar{a}_i \subset \overline{P}$ of the segments $\bar{\ell}_{n+i-3}$ $(i=1,\ldots,g+1)$ cut from the surface $\overline{P}$ a connected component $\overline{U}_{n-2} \subset \overline{P} - \bigcup_{i=1}^{g+1} \bar{a}_i$, not containing the contours of the segments $\bar{\ell}_i$ $(i=1,\ldots,n-3)$. Since $\ell \cap B_f = \varnothing$, there is a homotopy of the segments $\bar{\ell}_i$ to segments $\ell'_i$ $(i=1,\ldots,g+n-2)$ possessing the same properties (1)–(3) as the segments $\bar{\ell}_i$, but such that $\ell_a \notin U_{\ell'_i} \subset \overline{P}$ $(i=1,\ldots,n-2)$, where $U_{\ell'_{n-2}}$ is the image of the domain $\overline{U}_{n-2}$ under the deformation $\bar{\ell}_i \to \ell'_i$. But then the set of segments

$$\{\tilde{\ell}_1,\ldots,\tilde{\ell}_{g+n-1}\} = \{\ell, \ell'_1,\ldots,\ell'_{g+n-2}\}$$

satisfies conditions (1)–(3) of Lemma 1.6.

Deformations of the segments $\{\tilde{\ell}_i\}$ leaving them disjoint preserve properties (1)–(3). Using such deformations, one can easily satisfy condition (4) of Lemma 1.6.                                                                                                         □

THEOREM 1.1 ([**15**]). *Let $f_i\colon P_i \to S$ $(i=1,2)$ be two coverings with the same number $n$ of sheets by surfaces of the same genus $g$ having only simple critical points. Then there are homeomorphisms $\phi_p\colon P_1 \to P_2$ and $\phi_s\colon S \to S$ such that $f_2\phi_p = \phi_s f_1$.*

Proof. By Lemma 1.6, there are sets of vanishing segments $\ell_1^i,\ldots,\ell_{g+n-1}^i$ $\subset P_i$ $(i=1,2)$, satisfying assumptions (1)–(3) of the lemma. Let $\phi_s\colon S \to S$ be a homeomorphism such that

$$\phi_s(\ell_t^1) = \ell_t^2 \quad (t=1,\ldots,g+n-1).$$

Because of assumptions (1)–(3) of Lemma 1.6, the contours of the segments $\ell_t^i$ cut the surfaces $P_1$ and $P_2$ into one-sheeted domains $U_t^i$ $(i=1,2;\ t=1,\ldots,n)$ with the same, with respect to $t$, scheme of gluing the sheets. The equation $\phi_t = f_2^{-1}\phi_s f_1$ defines homeomorphisms of $U_t^1 \to U_t^2$ consistent on the boundaries of the sheets. Their union is a homeomorphism $\phi_p\colon P_1 \to P_2$ such that $f_2\phi_p = \phi_s f_1$.                                                                     □

**5.**     Let $f\colon P \to S$ be an $n$-sheeted covering, let $\ell \subset P$ be a one-sheeted segment with respect to $f$ and let $\tilde{\ell} = f(\ell)$. Consider the compactifications $\overline{P}$ and $\overline{S}$ of the surfaces $P - \ell$ and $S - \tilde{\ell}$ by closed contours $a_p \subset \overline{P}$ and $a_s \subset \overline{S}$. Let $\phi_p\colon \overline{P} \to P$ and $\phi_s\colon \overline{S} \to S$ be the natural continuous mappings such that

$$\phi_p|_{\overline{P}-a_p}\colon (\overline{P} - a_p) \to (P - \ell) \quad \text{and} \quad \phi_s|_{\overline{S}-a_s}\colon (\overline{S} - a_s) \to (P - \tilde{\ell})$$

are one-to-one and

$$\phi_p|_{a_p}\colon a_p \to \ell \quad \text{and} \quad \phi_s|_{a_s}\colon a_s \to \tilde{\ell}$$

are two-sheeted coverings. Then there is a homeomorphism $\phi_a\colon a_p \to a_s$ such that $f\phi_p|_{a_p} = \phi_s\phi_a$. Denote by $\tilde{P}$ the surface obtained from $P \cup S$ by identification of the contours $a_p$ and $a_s$ by means of the homeomorphism

$\phi_a$. The coverings $f\phi_p$ and $\phi_s$ induce an $(n+1)$-covering $\tilde{f}\colon \tilde{P}\to S$, which we call the *extension of the covering $f$ along the segment $\ell$.*

THEOREM 1.2. *For any $g\geqslant 0$ and $n>1$ there is an $n$-sheeted covering $f\colon P\to S$ by a surface $P$ of genus $g$ having only simple critical points.*

Proof. The proof proceeds by induction on $n$. For $n=2$ this is the well-known hyperelliptic covering. Suppose the theorem is proved for $n<d$ and $n=d>2$. Consider an $(n-1)$-sheeted covering $f'\colon P'\to S$ by a surface of genus $g$ having only simple critical points. Let $\ell\subset P'$ be a one-sheeted segment with respect to $f'$ such that $f'(\ell)\cap B_{f'}=\varnothing$. Then the extension $f\colon P\to S$ of the covering $f'$ along the segment $\ell$ is an $n$-sheeted covering having only simple ramifications.                                  □

## 2. Coverings with a single degenerate critical value

**1.** We prove two lemmas.

LEMMA 2.1. *Let $s_0\in B_f$ and suppose the set $f^{-1}(s_0)$ consists of a single point $p_0$, and all critical values in the set $B_f-\{s_0\}$ are simple. Then there is a set of vanishing segments $\ell_1,\dots,\ell_{g+n-1}$ such that: (1) $\ell_i\cap\ell_j=\varnothing$ if $i>n-1$ or $j>n-1$; (2) $\ell_i\cap\ell_j=s_0$ for $i\leqslant n-1$, $j\leqslant n-1$; (3) $\ell_1,\dots,\ell_{n-2}$ are cutting segments; (4) the union of contours $a_i$ of the segments $\ell_{n+i-2}$ $(i=1,\dots,g+1)$ cuts from $P$ a sphere with holes $U_{n-1}\subset P-\bigcup_{i=1}^{g+1}a_i$ that does not contain the contours of the segments $\ell_i$ $(i=1,\dots,n-2)$.*

Proof. The proof proceeds by induction on $n$. For $n=2$ the assertion of the lemma is obvious. Suppose the lemma is proved for $n<d$ and let $n=d>2$. Then, by Lemma 1.5, there is a cutting segment $\ell$ containing the point $s_0=f(p_0)$. Let $\bar{f}\colon \overline{P}\to S$ be the reduction with respect to this segment, let $a$ denote its contour, and let $\ell_a\subset P$ denote the trace of the reduction. By the induction hypothesis, there are vanishing segments $\bar{\ell}_1,\dots,\bar{\ell}_{g+n-2}$ with respect to $f$ such that (1) $\bar{\ell}_i\cap\bar{\ell}_j=\varnothing$ if $i>n-2$ or $j>n-2$; (2) $\bar{\ell}_i\cap\bar{\ell}_j=s_0$ for $i\leqslant n-2$, $j\leqslant n-2$; (3) $\bar{\ell}_1,\dots,\bar{\ell}_{n-3}$ are cutting segments with respect to $\bar{f}$; (4) the union of the contours $\bar{a}_i\subset\overline{P}$ of the segments $\bar{\ell}_{n+i-3}$ $(i=1,\dots,g+1)$ cuts from the surface $\overline{P}$ the sphere with holes

$$\overline{U}_{n-2}\subset\overline{P}-\bigcup_{i=1}^{g+1}\bar{a}_i$$

that does not contain the contours of the segments $\bar{\ell}_i$ $(i=1,\dots,n-2)$.

Since one of the ends of the segment $\ell$ does not belong to the set $B_f$, there is a homotopy taking the segments $\bar{\ell}_i$ to segments $\ell_i$ with the same properties (1)–(4) as the segments $\bar{\ell}_i$ but not intersecting the set $\ell-\{s_0\}$. Under this homotopy, the component $\overline{U}_{n-2}$ is taken to $U_{n-2}$. Denote by $U_i=U_{\ell_i}\subset\overline{P}$ $(i=1,\dots,n-3)$ the films of the segments $\ell_i$ with respect to the mapping $\bar{f}$. If $\ell_a\notin\bigcup_{i=1}^{n-2}U_i$, then $\ell,\ell_1,\dots,\ell_{g+n-2}$ is the desired set of segments. Suppose that $\ell_a\subset U_t$. By a homotopy in the set $S-\ell_t$ we can take the segment $\ell$ to a segment $\tilde{\ell}$ such that $\tilde{\ell} \cap \ell_i = \varnothing$ $(i = 1, \ldots, g + n - 2)$, and there are no segments of the form $\ell_i$ between the segments $\tilde{\ell}$ and $\ell_t$ (see Fig. 3.2.1). Under the homotopy, the segment $\ell_a \in \overline{P}$ is transformed to the segment $\tilde{\ell}_a$ inside the film $U_t$. Therefore, $\tilde{\ell}$ is a cutting segment with respect to $\bar{f}$ and $\tilde{\ell}_a$ is the trace of the reduction modulo the contour $\tilde{a}$ of the segment $\tilde{\ell}$. The deformation in the set $S - B_{\bar{f}}$ of the segment $\ell_t$ to the segment $\ell_t'$ shown in Fig. 3.2.1 takes the film $U_t$ to the film $U' \subset \overline{P}$ not containing the segment $\tilde{\ell}_a$. Therefore, $\tilde{\ell}, \tilde{\ell}_t, \ell_i$ $(i \neq t)$ becomes the desired set of segments after a renumbering. $\qquad\square$

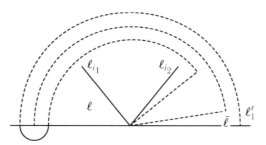

FIGURE 3.2.1

LEMMA 2.2. *Let* $s_0 \in S$, $f^{-1}(s_0) = \{p_1, \ldots, p_m\}$, *where* $m > 1$ *and* $k_i$ *is the ramification order of* $f$ *at* $p_i$ $(i = 1, \ldots, m)$, *and suppose all critical values in the set* $B_f - s_0$ *are simple. Then there are* $g + n - 1$ *vanishing segments* $\ell_{1,1}, \ldots, \ell_{1,k-1}, \ell_{2,1}, \ldots, \ell_{2,k_2-1}, \ell_{3,0}, \ldots, \ell_{3,k_3-1}, \ldots, \ell_{m,0}, \ldots, \ell_{m,k_m-1},$ $\ell_{m+1,0}, \ldots, \ell_{g+m-1,0}$ *such that* (1) $\ell_{i_1,j_1} \cap \ell_{i_2,j_2} = \varnothing$ *if* $j_1 \cdot j_2 = 0$; (2) $s_0 = \ell_{i_1,j_1} \cap \ell_{i_2,j_2}$ *if* $j_1 \cdot j_2 > 0$; (3) *in a neighborhood of* $s_0$ *the segments* $\ell_{i,j}$ *are situated as shown in Fig. 3.2.2;* (4) *for* $j > 0$ $\ell_{i,j}$ *is a cutting segment;* (5) *the contours* $a_{i_1,j_1}$ *and* $a_{i_2,j_2}$ *of the segments* $\ell_{i_1,j_1}$ *and* $\ell_{i_2,j_2}$ *intersect each other if and only if* $i_1 = i_2$ *and* $j_1 \cdot j_2 > 0$, *and in this case* $a_{i_1,j_1} \cap a_{i_2,j_2} = p_i$; (6) *for* $3 \leqslant i \leqslant m$ *the contour* $a_{i,0}$ *of the segment* $\ell_{i,0}$ *cuts from the surface* $P$ *a sphere with the hole* $U_i$ *such that* $a_{t,j} \subset U_i$ *if and only if* $t = i$, $j > 0$; (7) *the contours* $a_{m+1,0}, \ldots, a_{m+g-1,0}$ *of the segments* $\ell_{m+1,0}, \ldots, \ell_{m+g-1,0}$ *cut the surface* $P$ *into two spheres with holes, the first of which, denoted* $U_1$, *contains all contours of the form* $a_{1,j}$, *and the second,* $U_2$, *contains all contours of the form* $a_{i,j}$, *where* $2 \leqslant i \leqslant m$.

Proof. The proof proceeds by induction on $k = \sum_{i=1}^{m}(k_i - 1)$. For $k = 0$ the assertion of Lemma 2.2 follows from Lemma 1.6. Suppose Lemma 2.2 is proved for $k < d$ and $k = d > 0$. Then, by Lemma 1.5, there is a cutting segment $\ell \ni s_0$. Let $a$ be its contour and let $p_t = a \cap f^{-1}(s_0)$. Consider the reduction $\bar{f} : \overline{P} \to S$ of $f$ modulo $\ell$, and let $\ell_a \subset \overline{P}$ be the trace of the reduction. Applying the induction hypothesis, we find a set of segments $\{\bar{\ell}_{i,j}\}$ vanishing with respect to $\bar{f}$ and satisfying all requirements of Lemma 2.2. The contours of the segments $\bar{\ell}_{0,j}$ cut the surface $\overline{P}$ into spheres with holes

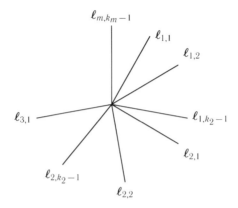

FIGURE 3.2.2

$U_i \ni \bar{\ell}_{i,j} \ni p_i$ $(j > 0)$. Since $p_t \subset \ell_a$ and $\bar{f}(\ell_a - p_t) \subset S - B_f$, there is a deformation in the set $S - B_{\bar{f}}$ taking the set of segments $\ell_{i,j}$ to the set of segments $\tilde{\ell}_{i,j}$ such that the segment $\ell_a$ belongs to the set $\tilde{U}_t$ cut off by the contour of the segment $\tilde{\ell}_{t,0}$, but does not belong to the set of films of the segments of the form $\tilde{\ell}_{t,j}$ $(j > 0)$. This deformation can be chosen in such a way that after an appropriate renumbering of the segments $\ell$ and $\tilde{\ell}_{t,j}$ we obtain the desired set of segments. $\qquad\square$

**2.** The following theorem holds.

THEOREM 2.1. (1) *Let* $g \geqslant 0, n > 1$, $s_0 \in S$ *and let* $\ell_1, \ldots, \ell_{g+n-1} \subset S$ *be a set of segments such that:* (a) $\ell_i \cap \ell_j = \varnothing$ *if* $i > n - 1$ *or* $j > n - 1$; (b) $\ell_i \cap \ell_j = s_0$ *for* $i \leqslant n - 1$, $j \leqslant n - 1$. *Then there is a compact surface* $P$ *of genus* $g$ *and an* $n$-*sheeted covering* $f : P \to S$ *such that the set* $f^{-1}(s_0)$ *consists of a single point* $p_0 \in A_f$, *all critical values in the set* $B_f - \{s_0\}$ *are simple, and the segments* $\ell_i$ $(i = 1, \ldots, g + n - 1)$ *form a set of segments vanishing with respect to* $f$ *and satisfying requirements* (1)–(4) *of Lemma 2.1.* (2) *Let* $g \geqslant 0$, $n > 1$, $s_0 \in S$, $m > 1$, $n = \sum_{i=1}^m k_i$ (*where* $k_i > 0$) *and let* $\ell_{1,1}, \ldots, \ell_{1,k_1-1}$, $\ell_{2,1}, \ldots, \ell_{g+m-1,0} \subset S$ *be a set of segments such that* (a) $\ell_{i_1,j_1} \cap \ell_{i_2,j_2} = \varnothing$ *if* $j_1 \cdot j_2 = 0$; (b) $\ell_{i_1,j_1} \cap \ell_{i_2,j_2} = s_0$ *if* $j_1 \cdot j_2 > 0$; (c) *in a neighborhood of* $s_0$ *the segments* $\ell_{i,j}$ *are situated as shown in Fig. 3.2.2. Then there is a compact surface* $P$ *of genus* $g$ *and an* $n$-*sheeted covering* $f : P \to S$ *such that all critical points in the set* $B_f - S_0$ *are simple,* $f^{-1}(s_0) = \{p_1, \ldots, p_m\}$, *and the ramification order of* $f$ *at the point* $p_i$ *is* $k_i$, *and the segments* $\ell_{i,j}$ $(i = 1, \ldots, g + m - 1$, $j = 0, \ldots, k_i)$ *form a set of segments vanishing with respect to* $f$ *and satisfying requirements* (1)–(7) *of Lemma 2.2.*

PROOF. (1) Let $f_0 : P_0 \to S$ be a two-sheeted covering by a surface $P_0$ of genus $g$ such that $s_0 \in B_{f_0}$. Then the desired covering $f : P \to S$ can be obtained by successively extending the covering $f_0$ along the segments $\ell_1, \ldots, \ell_{g+n-i}$.

(2) Let $f_0\colon P_0 \to S$ be a two-sheeted covering by a surface $P_0$ of genus $g$ such that $s_0 \notin S - B_f$. By extending along the segments $\ell_{s,0}, \ldots, \ell_{g+m-1,0}$, construct an $m$-sheeted covering $f_1\colon P_1 \to S$. Let $f_1^{-1}(s_0) = \{p_1, \ldots, p_m\}$ and let $\tilde{\ell}_{i,j}$ be the connected component of the preimage $f_1^{-1}(\ell_{i,j})$ $(i = 1, \ldots, m,$ $j = 1, \ldots, k_i - 1)$ containing the point $p_i$. Then the desired covering $f\colon P \to S$ can be constructed by successively extending the covering $f_1\colon P_1 \to S$ along the segments $\tilde{\ell}_{i,j}$ $(i = 1, \ldots, m, j = 1, \ldots, k-1)$.                $\square$

COROLLARY 2.1. *Let* $s_0 \subset S$, $g \geqslant 0$, $n \geqslant 1$ *and let* $n = \sum_{i=1}^{m} k_i$, *where* $k_i > 0$, $m > 0$. *Then there is an* $n$-*sheeted covering* $f\colon P \to S$ *of the sphere* $S$ *by a surface* $P$ *of genus* $g$ *such that all the critical values in the set* $B_f - \{s_0\}$ *are simple and the set* $f^{-1}(s_0)$ *decomposes into* $m$ *points where the mapping* $f$ *has ramification orders* $k_1, \ldots, k_m$.

**3.**   We need three more lemmas.

LEMMA 2.3. *Let* $g = 0, n = 3$ *and* $p_1, p_2 \subset A_f$, *and suppose* $p_2$ *is a simple critical point. Then there is a cutting segment* $\ell$ *whose contour* $a_\ell$ *contains the points* $p_1$ *and* $p_2$.

Proof. Set $s_i = f(p_i)$ $(i = 1, 2)$. If $s_1$ is a degenerate critical value, then the assertion of Lemma 2.3 follows from Lemma 2.1. Let $s_1$ be a simple critical value. Then, by Lemma 1.6, there are disjoint cutting segments $\ell_1$ and $\ell_2$. If $s_i \in \ell_i$ $(i = 1, 2)$, then the segment $\ell$ shown in Fig. 3.2.3 also is a cutting segment.                $\square$

LEMMA 2.4. *Let* $n > 2$ *and* $p_1, p_2 \in A_f$, *and suppose the set* $B_f - \{f(p_1)\}$ *consists of simple critical values. Then there is a cutting segment* $\ell$ *whose contour contains the points* $p_1$ *and* $p_2$.

Proof. We proceed by induction on $g + n$. For $g + n = 3$ the assertion of Lemma 2.4 follows from Lemma 2.3. Suppose Lemma 2.4 is proved for $g + n < d$, let $g + n = d > 3$ and suppose there is no cutting segment whose contour contains the points $p_1$ and $p_2$. Then, by Lemmas 2.1 and 2.2, there is a cutting segment $\ell^0$ whose contour does not contain the points $p_1$, $p_2$ and is either a cutting segment or a segment whose contour does not cut the surface $P$.

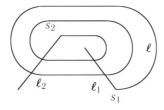

FIGURE 3.2.3

Let $\ell^0$ be a cutting segment, let $\bar{f}\colon \overline{P} \to S$ be the reduction modulo this segment, let $a^0$ be its contour, and let $\ell_a^0 \subset \overline{P}$ be the trace of the reduction.

By the induction hypothesis, there is a cutting segment $\bar{\ell} \subset S$ with respect to $\bar{f}$ whose contour $\bar{a} \subset \overline{P}$ contains the points $p_1$ and $p_2$. Since $\ell_a^0 \cap B_{\bar{f}} \subset f(p_1)$, there is a deformation in the set $S - B_f$ taking the segment $\bar{\ell}$ to the segment $\ell$ whose film does not contain the trace $\ell_a^0$ of the reduction. In this case $\ell$ is a cutting segment with respect to $f$ whose contour contains the points $p_1$ and $p_2$.

Now let $f(p_1) \notin f(a^0)$, let $P - a^0$ be a connected set, let $\bar{f} \colon \overline{P} \to S$ be the reduction modulo the contour $a^0$, and let $\ell^1, \ell^2 \subset \overline{P}$ be its trace. By the induction hypothesis, there is a cutting segment $\bar{\ell} \subset S$ with respect to $\bar{f}$ whose contour $\bar{a} \subset \overline{P}$ contains the points $p_1$ and $p_2$. To complete the proof, it suffices to show that there is a deformation in the set $S - B_f$ taking the segment $\bar{\ell}$ to a segment $\ell$ whose film does not contain the trace $\ell^1 \cup \ell^2$ of the reduction. The rest of the argument repeats word for word the corresponding part of the proof of Lemma 1.5. $\qquad\square$

LEMMA 2.5. *Suppose all critical points of a mapping $f$ are simple, $B_f = \{a_1, b_1, \ldots, a_{g+n-1} b_{g+n-1}\}$; $s_1, \ldots, s_m \in S - B_f$ and $f^{-1}(s_k) = \{p_1^k, \ldots, p_n^k\}$ $(k = 1, \ldots, m)$. Then there is a set of cutting segments $\ell_1, \ldots, \ell_{g+n-1}$ satisfying all properties of Lemma 1.6 and such that $a_i$, $b_i$ are the ends of the segment $\ell_i$ $(i = 1, \ldots, g + n - 1)$.*

The proof of Lemma 2.5 is similar to that of Lemma 1.6, the only difference being the replacement of Lemma 1.5 by Lemma 2.4.

**4.** The following theorem holds.

THEOREM 2.2. *Let $f_i \colon P_i \to S$ $(i = 1, 2)$ be two coverings with the same number $n$ of sheets by surfaces of the same genus $g$. Let $s_i \in S$ $(i = 1, 2)$ be two points such that the sets $B_{f_i} - \{s_i\}$ consist of simple critical values of the coverings $f_i$. Let $A_i = f_i^{-1}(s_i) \cup A_{f_i}$ and let $\phi \colon A_1 \to A_2$ be a one-to-one correspondence such that $\phi(f_1^{-1}(s_1)) = f_2^{-1}(s_2)$ and for any point $p \in A_1$ the ramification order of $f_1$ at $p$ coincides with the ramification order of $f_2$ at $\phi(p)$. Then there are homeomorphisms $\phi_P \colon P_1 \to P_2$ and $\phi_S \colon S \to S$ such that $\phi_P|_{A_1} = \phi$ and $\phi_S f_1 = f_2 \phi_P$.*

Proof. We proceed by induction on $n$. For $n = 2$ the assertion of Theorem 2.2 is obvious. Now suppose the theorem is proved for $n < d$ and $n = d > 3$. Then, by Lemma 2.4, there are cutting segments $\ell_i$ with respect to $f_i$ whose contours $a_i$ contain the points $p_i', p_i'' \in A_{f_i}$ and such that $f_i(p_i') = s_i$, $\phi(p_1') = p_2'$ and $\phi(p_1'') = p_2''$. Let $\bar{f}_i \colon \overline{P}_i \to S$ be the reduction modulo these segments and let $\ell_a^i \subset P$ be the traces of these reductions. Because of the inclusion $p_i' \in \ell_a^i$ together with the induction hypothesis, there are homeomorphisms $\phi_S \colon S \to S$ and $\phi_{\overline{P}} \colon \overline{P} \to \overline{P}$ such that $\phi_S \bar{f}_1 = \bar{f}_2 \phi_{\overline{P}}$, $\phi_{\overline{P}}|_{A_1 - p_1''} = \phi_{\overline{P}}|_{\bar{f}^{-1}(s_1) \cup A_{\bar{f}_1}} = \phi|_{A_1 - p_1''}$ and $\phi_{\overline{P}}(\ell_a^1) = \ell_a^2$. Since the films $U_{\ell_i}$ are one-sheeted, the mapping $\phi_{\overline{P}}$ admits a unique extension as a homeomorphism $\phi_P \colon P_1 \to P_2$ satisfying all requirements of Theorem 2.2. $\qquad\square$

COROLLARY 2.2. *Let $f\colon P \to S$ be an $n$-sheeted covering by a surface $P$ of genus $g$, let $s_0 \in S$, and suppose the set $B_f - \{s_0\}$ consists of simple critical values, while the set $f^{-1}(s_0)$ consists of points $p_0, p_1, \ldots, p_m$, where $p_0 \in A_f$. Then there are pairwise disjoint segments $\ell_1, \ldots, \ell_m$ vanishing with respect to $f$ and such that for all $i$ the contour $\ell_i$ cuts from $P$ a domain $U_i$ homeomorphic to the sphere and such that $U_i \cap f^{-1}(s_0) = p_i$.*

Proof. Let $k_0, k_1, \ldots, k_m$ be the ramification orders of $f$ at the points $p_0, \ldots, p_m$, and let $s_1, \ldots, s_q$ be the critical values of $f$ other than $s_0$. Denote by $f_0\colon P_0 \to S$ the two-sheeted covering with the critical values $s_0, s_1, \ldots,$ $s_{2g+1}$. Set $\tilde{p}_0 = f_0^{-1}(s_0)$. Let $a_1, \ldots, a_m \subset S$ be pairwise disjoint segments with the ends $s_{2g+2}, s_{2g+3}, \ldots, s_{2g+m+1}$, let $\tilde{a}_i \subset P_0$ be a connected component of the preimage $f_0^{-1}(a_i)$, and let $f_1\colon P_1 \to S$ be the extension of the covering $f_0\colon P_0 \to S$ along the segments $\tilde{a}_1, \ldots, \tilde{a}_m$. Let $\tilde{U}_i \subset P_1$ be the film of the segment $a_i$ and let $\tilde{p}_i = f^{-1}(s_0) \cap \tilde{U}_i$. Denote by $d_{1,1}, \ldots, d_{m,1}, \ldots, d_{m,k_m-1} \subset S$ pairwise disjoint segments not intersecting the segments $a_1, \ldots, a_m$ and connecting the point $s_0$ to $s_{2(g+m+1)}, \ldots, s_q$. Set $d_{i,j} = f_1^{-1}(d_{i,j}) \cap U_i$ and denote by $\tilde{f}\colon \tilde{P} \to S$ the extension of the covering $\tilde{f}$ along the segments $\tilde{d}_{i,j}$.

By Theorem 2.2, there are homeomorphisms $\phi_P\colon \tilde{P} \to P$ and $\phi_S\colon S \to S$ such that $\phi_P(\tilde{p}_i) = p_i$ and $\phi_S \tilde{f} = f \phi_P$. The segments $\ell_i = \phi_S(a_i)$ satisfy all the requirements of Corollary 2.2. $\qquad\square$

## 3. Spaces of complex meromorphic functions

**1.**    If not specified otherwise, $\Gamma$ denotes a compact Riemann surface of genus $g$, $S = \mathbb{C} \cup \infty$ is the Riemann sphere and $f\colon P \to S$ is an $n$-sheeted holomorphic mapping. A pair $(P, f)$ is called an *$n$-sheeted meromorphic function of genus $g$*. Two meromorphic functions $(P_1, f_1)$ and $(P_2, f_2)$ are considered to be the same if there is a biholomorphic mapping $\phi\colon P_1 \to P_2$ such that $f_1 = f_2 \phi$.

Denote by $H_{g,n}^{\mathbb{C}}$ the set of all $n$-sheeted meromorphic functions of genus $g$. In other words, $H_{g,n}^{\mathbb{C}}$ is the set of all meromorphic functions of degree $n$ on nonsingular irreducible complex algebraic curves of genus $g$.

Let $P_0$ be a compact orientable surface of genus $g$ and let $\tilde{H}_{g,n}^{\mathbb{C}}$ be the space of all $n$-sheeted coverings $f\colon P_0 \to S$ endowed with the compact-open topology. Two coverings $f_1\colon P_0 \to S$ and $f_2\colon P_0 \to S$ are said to be equivalent if there is a homeomorphism $\phi\colon P_0 \to P_0$ such that $f_1 = f_2 \phi$. Denote by $\overline{H}_{g,n}^{\mathbb{C}}$ the topological quotient space of $\tilde{H}_{g,n}^{\mathbb{C}}$ modulo this equivalence relation.

According to the Kerekjarto theorem [39], for any $n$-sheeted covering $f\colon P_0 \to S$ there is a unique complex structure on $P_0$ with respect to which $f$ is a holomorphic mapping. This property allows one to establish a one-to-one correspondence between the sets $\overline{H}_{g,n}^{\mathbb{C}}$ and $H_{g,n}^{\mathbb{C}}$ and in this way introduce a topology on $H_{g,n}^{\mathbb{C}}$.

**2.**   The following lemma holds.

LEMMA 3.1. *Let $U$ and $V$ be two sets homeomorphic to the closed disk $\Lambda = \{z \in \mathbb{C} \,|\, |z| \leqslant 1\}$, let $\varphi U$ and $\varphi V$ denote their boundaries, and let $\varphi\phi\colon \varphi V \to \varphi U$ be an n-sheeted covering. Then for any points $p_1, \ldots, p_{n-1} \subset U - \varphi U$ ($p_i \neq p_j$ for $i \neq j$) there is an n-sheeted covering $\phi\colon V \to U$ with simple critical values $p_1, \ldots, p_{n-1}$ such that $\phi|_{\varphi V} = \varphi\phi$.*

Proof. Let $f(\omega) = \omega^n + a_{n-1}\omega^{n-1} + \cdots + a_0$ be a polynomial with pairwise distinct critical values $z_i$ such that $|z_i| < 1$ ($i = 1, \ldots, n-1$). Then the set $\tilde{V} = f^{-1}(\Lambda)$ is homeomorphic to $\Lambda$, the set $\varphi\tilde{V} = f^{-1}(\varphi\Lambda)$, where $\varphi\Lambda = \{z \subset \mathbb{C} \,|\, |z| = 1\}$, is homeomorphic to a circle, and the mapping $f|_{\tilde{V}}\colon \tilde{V} \to \Lambda$ is $n$-sheeted. Therefore, there are homeomorphisms $\psi_U\colon \Lambda \to U$ and $\psi_V\colon \tilde{V} \to V$ such that $\psi_U(z_i) = p_i$ and $\varphi\phi = (\psi_U f \psi_V^{-1})|_{\varphi V}$. Set $\phi = \psi_U f \psi_V$.   □

**3.**   We need two more lemmas.

LEMMA 3.2. *Let $(P, f)$ be a meromorphic function, let $H_{g,n}^{\mathbb{C}} \supset W \supset (P, f)$ be its neighborhood and let $p_0 \in A_f$ be a degenerate ramification point of order $k$. Then there is a meromorphic function $(P', f') \subset W$ such that (1) $B_{f'} \cap B_f = B_f - f(p_0)$; (2) for any point $s \in B_f - f(p_0)$ there is a one-to-one correspondence $\phi\colon f^{-1}(s) \to (f')^{-1}(s)$ such that the ramification order of $f$ at a point $p \in f^{-1}(s)$ coincides with the ramification order of $f'$ at the point $\phi(p)$; (3) the set $B_{f'} - B_f$ consists of $k-1$ simple critical values.*

Proof. Let $U$ be a neighborhood of $p_0$ such that $f(U) \cap B_f = f(p_0)$. By applying Lemma 3.1, let us construct a continuous $n$-sheeted mapping $f'\colon P \to S$ coinciding with $f$ on $P - U$ and having exactly $(k-1)$ simple critical points on $U$. This mapping possesses properties (1)–(3). On the other hand, according to the Kerekjarto theorem, the surface $P$ can be endowed with a complex structure with respect to which $f'$ is a holomorphic mapping. Choosing the neighborhood $U$ sufficiently small one can obtain a meromorphic function $(P', f')$ in the neighborhood $W$.   □

COROLLARY 3.1. *The space $H_{g,n}^{\mathbb{C}}$ coincides with the closure of the space $H_{g,n}$ of meromorphic functions having only simple critical values.*

LEMMA 3.3. *Let $(P, f) \in H_{g,n}^{\mathbb{C}}$, let $s_1, \ldots, s_k \in B_f$ be simple critical values, and let $\ell_i\colon [0,1] \to S$ ($i = 1, \ldots, k$) be continuous mappings such that (1) $\ell_i(0) = s_i$; (2) $\ell_i([0,1]) \subset P$ is a segment; (3) the connected components $\tilde{\ell}_i$ of the preimage $f^{-1}(\ell_i[0,1])$ containing the point $p_i = f^{-1}(s_i) \cap A_f$ do not intersect each other; (4) for $t = (0,1]$ we have $\ell_i(t) \in S - B_f$. Then there is a continuous mapping $\Phi\colon [0,1] \to H_{g,n}^{\mathbb{C}}$ such that the meromorphic functions $(P_t, f_t) = \Phi(t)$ satisfy the following conditions: (1) $(P_0, f_0) = (P, f)$; (2) $B_{f_t} = \left( B_f - \bigcup_{i=1}^{k} s_i \right) \cup \left( \bigcup_{i=1}^{k} \ell_i(t) \right)$.*

Proof. By the assumption, there are pairwise disjoint neighborhoods $U_i(P)$ of the segments $\tilde{\ell}_i$ such that $U_i \cap A_f = p_i$. Applying Lemma 3.1 to each $t \in [0,1]$, we construct a continuous $n$-sheeted mapping $f_{t_k} : P \to S$ such that

$$f_t|_{P-\bigcup_{i=1}^k U_i} = f|_{P-\bigcup_{i=1}^k U_i} \quad \text{and} \quad B_{f_t} = \left( B_f - \bigcup_{i=1}^k s_i \right) \cup \left( \bigcup_{i=1}^k \ell_i(t) \right).$$

By the Kerekjarto theorem, the surface $P$ admits a unique complex structure $P_t$ with respect to which $f_t$ is holomorphic. The continuity of the constructed mapping $t \to (P_t, f_t)$ follows from the definition of the topology on $H_{g,n}^{\mathbb{C}}$. $\qquad \square$

**4.** Denote by

$$H_{g,n}^{\mathbb{C}}(r_1, \dots, r_k) \subset H_{g,n}^{\mathbb{C}}$$

the space of meromorphic functions $(P, f) \in H_{g,n}^{\mathbb{C}}$ whose divisor of poles $f^{-1}(\infty)$ contains pairwise distinct points $p_1, \dots, p_k$ of multiplicities $r_1, \dots, r_k$, respectively. Obviously,

$$H_{g,n}^{\mathbb{C}}(r_1, \dots, r_n) = \varnothing \quad \text{for} \quad \sum_{j=1}^n r_j > n.$$

THEOREM 3.1. *For $\sum_{j=1}^k r_j \leqslant n$ the space $H_{g,n}^{\mathbb{C}}(r_1, \dots, r_k)$ is nonempty, connected, and its real dimension is $4(g+n-1)-2\sum_{j=1}^k(r_j-1)$.*

Proof. By Lemma 3.2, the set $H_{g,n}^{\mathbb{C}}(r_1, \dots, r_k)$ is the closure of the space

$$\tilde{H}_{g,n}^{\mathbb{C}}(r_1, \dots, r_k) \subset H_{g,n}^{\mathbb{C}}(r_1, \dots, r_k)$$

consisting of meromorphic functions $(P, f) \in H_{g,n}^{\mathbb{C}}$ such that (1) the set $B_f \cap \mathbb{C}$ consists of simple critical values; (2) the divisor $f^{-1}(\infty)$ consists of points of multiplicities $r_1, \dots, r_k$ and $n - \sum_{j=1}^k r_i$ points of multiplicity 1.
Set

$$m = n - \sum_{j=1}^k (r_i - 1) \quad \text{and} \quad t = 2(g + n - 1) - \sum_{j=1}^k (r_j - 1).$$

By Corollary 2.1, for any set of points $s_1, \dots, s_t \subset \mathbb{C} = S - \infty$ there is a continuous $n$-sheeted covering $f : P_0 \to S$ by a surface $P_0$ of genus $g$ such that (1) $B_f = \bigcup_{i=0}^t s_i$, where $s_0 = \infty$; (2) the set $f^{-1}(s_0)$ decomposes into $m$ points $p_1, \dots, p_m$; (3) the ramification order of $f$ at the point $p_i$ is $r_i$ for $i \leqslant k$ and is 1 for $i > k$. By the Kerekjarto theorem, there is a unique complex structure on $P_0$ with respect to which $f$ is a holomorphic mapping. The resulting meromorphic function $(P, f)$ belongs to the set $\tilde{H}_{g,n}^{\mathbb{C}}(r_1, \dots, r_k)$ and, therefore, $\tilde{H}_{g,n}^{\mathbb{C}} \neq \varnothing$. This construction generates all functions in $\tilde{H}_{g,n}^{\mathbb{C}}(r_1, \dots, r_k)$. The points $s_1, \dots, s_t \subset \mathbb{C}$ are the continuous parameters of the construction. Thus the real dimension of $H_{g,n}^{\mathbb{C}}(r_1, \dots, r_k)$ is

$$2t = 4(g + n - 1) - 2 \sum_{j=1}^k (r_j - 1).$$

Let us prove that the space $\tilde{H}^{\mathbb{C}}_{g,n}(r_1,\ldots,r_k)$ is connected. Let

$$(P_i, f_i) \in \tilde{H}^{\mathbb{C}}_{g,n}(r_1,\ldots,r_k) \quad (i=1,2),$$

and let $p_{i,j}$ be a point of multiplicity $r_j$ of the divisor $f_i^{-1}(\infty)$. Then, by Lemmas 2.1 and 2.2, there are ordered sets $L_i$ of vanishing segments with respect to $f_i$ satisfying either assumptions (1)–(4) of Lemma 2.1 or assumptions (1)–(7) of Lemma 2.2. Because of these properties, the sets $L_1$ and $L_2$ are such that there is a continuous family of homeomorphisms $d_t\colon S \to S$ ($t \in [a,b]$) possessing the following properties: (1) $d_a = 1$; (2) $d_b(L_1) = L_2$; (3) $d_t(\infty) = \infty$ for all $t \in [a,b]$. One can split the segment $[a,b]$ into segments $[a,t_1], [t_1,t_2], \ldots, [t_k,b]$ such that the segments of the form $d_{[t_j,t_{j+1}]} s_i \subset S$ (where $s_i \in B_f - \{s_0\}$) are pairwise disjoint. Applying Lemma 3.3 to these segments, we find a continuous path $\Phi\colon [a,b] \to H^{\mathbb{C}}_{g,n}(r_1,\ldots,r_k)$. Put $(P_b, f_b) = \Phi(b)$.

The sets $f_b^{-1}(L_2) = f_b^{-1}(d_b(L_1))$ and $f_2^{-1}(L_2)$ cut the surfaces $P_b$ and $P_2$ into 1-sheeted components with the same scheme of gluing the sheets. (This scheme is totally determined by the assumptions of Lemmas 2.1 and 2.2 concerning the sets of segments $L_1$ and $L_2$.) Hence, there is a biholomorphic mapping $\phi\colon P_b \to P_2$ such that $f_b = f_2\phi$ and, therefore, $(P_b, f_b) = (P_2, f_2)$. $\square$

REMARK 3.1. It follows from the proof of Theorem 3.1 that a general point of the space $H^{\mathbb{C}}_{g,n}(r_1,\ldots,r_k)$ depends on the set of finite simple critical values. These values can be regarded as a complete coordinate system in a neighborhood of a general point thus producing a complex space structure on $H^{\mathbb{C}}_{g,n}(r_1,\ldots,r_k)$. The complex dimension of this space is

$$2(g + n - 1) - \sum_{j=1}^{k}(r_j - 1).$$

REMARK 3.2. Denote by $H^{+}_{g,n}(r_1,\ldots,r_k) \subset H^{\mathbb{C}}_{g,n}$ the space of all meromorphic functions $(P,f) \in H^{\mathbb{C}}_{h,n}$ such that $B_f \cap \{z \subset \mathbb{C} \,|\, \operatorname{Im} z \leqslant 0\} \subset (-i)$ and the preimage $f^{-1}(-i)$ decomposes into points $p_1,\ldots,p_k$ of multiplicity $r_1,\ldots,r_k$. An easy modification of the proof of Theorem 3.1 shows that the space $H^{+}_{g,n}(r_1,\ldots,r_k)$ is connected. We use this statement in Section 5.

**5.** In the topological field theory [19], the space $H_g(r_1,\ldots,r_k)$ consisting of meromorphic functions with simple finite critical values and divisor of poles of the form $r_1 p_1 + \cdots + r_k p_k$ plays an important role. The automorphism group of the complex plane

$$\operatorname{Aut}(\mathbb{C}) = \{h^{A,B} \,|\, h^{A,B}(z) = Az + B, A \in \mathbb{C} \setminus 0, B \in \mathbb{C}\} \cong (\mathbb{C} \setminus 0) \times \mathbb{C}$$

acts on $H_g(r_1,\ldots,r_k)$ according to the following rule. Associate to a meromorphic function $(P,f) \in H_g(r_1,\ldots,r_k)$ with finite critical values $s_1,\ldots,s_k$ and an element $h^{A,B} \in \operatorname{Aut}(\mathbb{C})$ the set of segments $\ell_i\colon [0,1] \to S$, where

$$\ell_i(t) = (1 + t(A-1))s_i + tB.$$

By Lemma 3.3, the set $\{\ell_i\}$ determines a family of meromorphic functions $(P_t, f_t) \in H_g(r_1, \ldots, r_k)$. Set

$$h^{A,B}(P, f) = (P_1, f_1)$$

and

$$\widehat{H}_g(r_1, \ldots, r_k) = H_g(r_1, \ldots, r_k)/\operatorname{Aut}(\mathbb{C}).$$

Theorem 3.1 of this chapter and Theorem 6.1 of Chapter 1 immediately imply

THEOREM 3.2. *The space* $\widehat{H}_g(r_1, \ldots, r_k)$ *is homeomorphic to* $T/\operatorname{Mod}$, *where the space* $T$ *is homeomorphic to* $\mathbb{R}^{4(g+n-2)-2\sum_{j=1}^{k}(r_j-1)}$, *and* Mod *is a discrete group.*

For the case of Laurent polynomials (rational functions of genus 0 with the divisor of poles consisting of two points) this result was obtained earlier by V. I. Arnold [**3**] by means of involved methods of singularity theory.

## 4. Topological structure of real meromorphic functions

**1.**     A *real meromorphic function* on a real algebraic curve $(P, \tau)$ is a meromorphic function $f\colon P \to S$ such that $f(\tau p) = \overline{f(p)}$ for all $p \in P$. Two real meromorphic functions $(P_1, \tau_1, f_1)$ and $(P_2, \tau_2, f_2)$ are considered to be the same if there is a biholomorphic mapping $\phi\colon P_1 \to P_2$ such that $f_1 = f_2\phi$ and $\phi\tau_1 = \tau_2\phi$.

The definition of a real meromorphic function axiomatizes properties of a real algebraic function, that is, a function $f = f(z)$ defined by an equation

$$a_0(z)f^n + a_1(z)f^{n-1} + \cdots + a_n(z) = 0$$

whose coefficients $a_i(z)$ are polynomials with real coefficients. The involution $(f, z) \to (\bar{f}, \bar{z})$ induces an antiholomorphic involution $\tau\colon P \to P$ of the Riemann surface $P$ of the function $f(z)$ and $f(\tau p) = \overline{f(p)}$. It is easy to show that any real meromorphic function can be constructed in this way.

Denote by $H_{g,n}^{\mathbb{R}}$ the set of all real meromorphic functions $(P, \tau, f)$, where $P$ is a surface of genus $g$ and $f$ is an $n$-sheeted mapping. In order to introduce a topology on the set $H_{g,n}^{\mathbb{R}}$ we need the following

THEOREM 4.1. *Let* $\tau_1\colon P_1 \to P_1$ *be an involutive homeomorphism of a compact surface* $P_1$ *inverting the orientation, let* $\tau_2\colon P_2 \to P_2$ *be an antiholomorphic involution of a compact Riemann surface* $P_2$, *and let* $f\colon P_1 \to P_2$ *be a continuous* $n$-*sheeted covering (maybe, ramified) such that* $f\tau_1 = \tau_2 f$. *Then the surface* $P_1$ *can be endowed with a complex structure with respect to which* $\tau_1$ *is an antiholomorphic involution and* $f$ *is a holomorphic covering.*

Proof. Associate to each point $s \in P_2$ a pair $(V_s, \psi_1)$ consisting of a neighborhood $V_s$ of $s$ and a biholomorphic mapping $\psi_s\colon V_s \to \Lambda_0 = \{z \in \mathbb{C} \mid |z| < 1\}$ such that (1) $\tau_2 V_s \cap V_s = \varnothing$ if $\tau_2 s \neq s$; (2) $\tau_2 V_s = V_s$, $\psi_s(\tau_1\tilde{s}) = \overline{\psi_s(\tilde{s})}$ if $\tau_2 s = s$, $\tilde{s} \in V_s$; (3) the set $V_s \cap B_f$ either is empty or coincides with the point $s$. For

$p \in P_1$ define $U_p$ as the connected component of the preimage $f^{-1}(V_{f(p)})$ containing $p$. Then $U_p$ is a neighborhood of $p$ homeomorphic to $\Lambda_0$. According to the Kerekjarto theorem [39] the covering group $H_p$ of the covering $f|U_p\colon U_p \to V_{f(p)}$ is isomorphic to $\mathbb{Z}_k$, where $k$ is the ramification order of $f$ at $p$ and the orbit $H_p(\tilde{p})$ coincides with $(f|U_p)^{-1}(f(\tilde{p}))$ for all $\tilde{p} \in U_p$. Therefore, if $\tau_1 p = p$, then $\tau_1 H_p \tau_1 = H_p$. According to a generalization of the Kerekjarto theorem [39] this property implies the existence of a homeomorphism $\tilde{\phi}_p\colon U_p \to \Lambda_0$ such that $\tilde{\phi}_p H_p \tilde{\phi}_p^{-1}$ is the group generated by the rotation $z \mapsto z \exp(2\pi i/k)$, and if $\tau_1 p = p$, then $\tilde{\phi}_p \tau_1 \tilde{\phi}_p^{-1}(z) = \bar{z}$ for all $z \in \Lambda_0$. Hence the mapping $\psi_{f(p)} f \phi_p^{-1}\colon \Lambda_0 \to \Lambda_0$ can be expressed as a composition $\eta_k \theta_p$, where $\eta_k(z) = z^k$ $(z \in \Lambda_0)$, and the homeomorphism $\theta_p\colon \Lambda_0 \to \Lambda_0$ is such that $\theta_p(z \exp(2\pi i/k)) = \theta_p(z) \exp(2\pi i/k)$ and $\theta_p(\bar{z}) = \overline{\theta_p(z)}$. Set $\phi_p = \theta_p \tilde{\phi}_p\colon U_p \to \Lambda_0$. Then $f\phi_p^{-1}\colon \Lambda_0 \to V_{f(p)}$ is a holomorphic mapping and $\phi_p(\tau_1 \tilde{p}) = \overline{\phi_p(\tilde{p})}$ if $\tau_1 p = p$ and $\tilde{p} \in U_p$. The set $\{(U_p, \phi_p) \mid p \in P\}$ is a complex structure on $P_1$ possessing all the properties stated in Theorem 4.1. $\qquad \square$

Now let $P_0$ be a compact orientable surface of genus $g$. Denote by $\tilde{H}^{\mathbb{R}}_{g,n}$ the set of pairs $(\tau, f)$, where $\tau\colon P_0 \to P_0$ is an involution inverting the orientation and $f\colon P_0 \to S$ is an $n$-sheeted covering and $f(\tau p) = \overline{f(p)}$ for all $p \in P$. Introduce a topology on the set $\tilde{H}^{\mathbb{R}}_{g,n}$ by assuming that a sequence $(\tau_n, f_n)$ tends to a pair $(\tau_0, f_0)$ if the sequences $\tau_n$ and $f_n$ tend to the mappings $\tau_0$ and $f_0$, respectively, in the compact-open topology of the corresponding spaces of mappings. We call two pairs $(\tau_1, f_1)$ and $(\tau_2, f_2)$ *equivalent* if there is a homeomorphism $\phi\colon P_0 \to P_0$ such that $f_1 = f_2\phi$ and $\phi\tau_1 = \tau_2\phi$. Denote by $\tilde{H}^{\mathbb{R}}_{g,n}$ the topological quotient space of $\tilde{H}^{\mathbb{R}}_{g,n}$ modulo this equivalence relation.

By Theorem 4.1, to a pair $(\tau, f) \in \tilde{H}^{\mathbb{R}}_{g,n}$ there corresponds a unique complex structure on $P_0$ with respect to which $f$ is a holomorphic and $\tau$ is an antiholomorphic mapping. This fact allows us to establish a one-to-one correspondence between the sets $\tilde{H}^{\mathbb{R}}_{g,n}$ and $H^{\mathbb{R}}_{g,n}$ and in this way introduce a topology on $H^{\mathbb{R}}_{g,n}$.

**2.** The following assertion holds.

LEMMA 4.1. *Let $U$ and $V$ be sets homeomorphic to the closed disk $\Lambda = \{z \in \mathbb{C} \mid |z| \leqslant 1\}$, and let $\tau_U\colon U \to U$ and $\tau_V\colon V \to V$ be orientation inverting involutive homeomorphisms. Let $\varphi U \subset U$, $\varphi V \subset V$ be the boundaries of the disks, and let $\varphi\phi\colon \varphi V \to \varphi U$ be an $n$-sheeted covering such that $\tau_U \varphi\phi = \varphi\phi\tau_V$. Then for any set of pairwise distinct points $\{p_1, \ldots, p_{n-1}\} \subset U \setminus \varphi U$ such that*

$$\tau_U\left(\bigcup_{i=1}^{n-1} p_i\right) = \bigcup_{i=1}^{n-1} p_i,$$

*there is an $n$-sheeted covering $\phi\colon V \to U$ with simple critical values and such that $\phi|_{\varphi V} = \varphi\phi$ and $\phi\tau_V = \tau_U\phi$.*

Proof. The set $\{p_1, \ldots, p_{n-1}\}$ consists of $k_1$ points invariant with respect to $\tau_U$ and $k_2$ pairs of points transposed under $\tau_U$. Let $f(\omega) = \omega^n + a_{n-1}\omega^{n-1} + \cdots + a_0$ be a polynomial with real coefficients all whose critical values $z_1, \ldots, z_{n-1}$ are pairwise distinct, $|z_i| < 1$, and suppose there are exactly $k_1$ real points among them. Then the set $\tilde{V} = f^{-1}(\Lambda)$ is homeomorphic to $\Lambda$ and the set $\varphi\tilde{V} = f^{-1}(\varphi\Lambda)$ is homeomorphic to the circle. The complex conjugation involution induces orientation reversing involutive homeomorphisms $\tau_{\tilde{V}} \colon \tilde{V} \to \tilde{V}$ and $\tau_\Lambda \colon \Lambda \to \Lambda$, and $f\tau_{\tilde{V}} = \tau_\Lambda f$.

Since all orientation reversing involutions of a disk are conjugate in the group of autohomeomorphisms of the disk [**39**] (that is, they are topologically equivalent), there is a homeomorphism $\psi_U \colon \Lambda \to U$ such that

$$\tau_U \psi_U = \psi_U \tau_\Lambda \quad \text{and} \quad \psi_U\left(\bigcup_{i=1}^{n-1} z_i\right) = \bigcup_{i=1}^{n-1} p_i.$$

The mappings $\varphi\psi = (\psi_U f)\big|_{\varphi\tilde{V}}$ and $\varphi\phi$ are $n$-sheeted coverings over $\varphi U$, and we have $\varphi\psi\tau_V = \tau_U\varphi\psi$ and $\varphi\phi\tau_V = \tau_U\varphi\phi$. Therefore, since the involutions $\tau_{\tilde{V}}$ and $\tau_U$ are topologically equivalent, there is a homeomorphism $\psi_V \colon \tilde{V} \to V$ such that $\varphi\psi = \varphi\phi(\psi_V\big|_{\varphi\tilde{V}})$ and $\psi_V\tau_{\tilde{V}} = \tau_V\psi_V$. We set $\phi = \psi_U f\psi_V^{-1}$.  $\square$

LEMMA 4.2. *Let* $(P, \tau, f) \in H_{g,n}^{\mathbb{R}}$, *let* $p \in P^\tau \cap A_f$, *and let* $U$ *be a neighborhood of* $f(p)$ *such that* (1) *the closure of* $U$ *is homeomorphic to the disk* $\Lambda$; (2) $\overline{U} = U$; (3) $U \cap B_f = f(p)$. *Then the closure of the connected component* $V$ *of the preimage* $f^{-1}(U)$ *containing* $p$ *is homeomorphic to the disk* $\Lambda$ *and* $\tau V = V$.

The proof is obvious.

LEMMA 4.3. *Let* $(P, \tau, f) \in H_{g,n}^{\mathbb{R}}$; *let* $p_1, p_2$ *be simple critical values of the covering* $f$, *let* $\tau(p_1 \cup p_2) = p_1 \cup p_2$, *and let* $U$ *be a neighborhood containing the points* $f(p_1)$ *and* $f(p_2)$ *and such that* (1) *the closure of* $U$ *is homeomorphic to* $\Lambda$; (2) $\overline{U} = U$; (3) $U \cap B_f = p_1 \cup p_2$; (4) *the points* $p_1$ *and* $p_2$ *belong to the same connected component* $V$ *of the preimage* $f^{-1}(U)$. *Then* $\tau V = V$ *and* $f|_V$ *is either a 2-sheeted mapping or a 3-sheeted mapping and the closure of* $V$ *is homeomorphic to the disk.*

Proof. The equality $\tau V = V$ follows from $\tau(p_1 \cup p_2) = p_1 \cup p_2$. By contracting each connected component of the boundaries of the sets $U$ and $V$ to a point we obtain an $\tilde{n}$-sheeted covering $\tilde{f} \colon \tilde{V} \to \tilde{U}$ of the sphere $\tilde{U}$ by a compact surface $\tilde{V}$ of genus $\tilde{g}$. The covering $\tilde{f}$ has at most three critical values two of which are simple, while the third, $s_0$, is the image of the boundary of $U$. Applying the Riemann–Hurwitz formula we conclude that

$$\tilde{g} - 1 = -\tilde{n} + \frac{1}{2}\sum_{i=1}^{m}(k_i - 1) + 1 = -\tilde{n} + \frac{1}{2}\tilde{n} - \frac{1}{2}m + 1,$$

where $m$ is the number of points in the set $\tilde{f}^{-1}(s_0)$. Hence we obtain $1 \leqslant m = 2\left(2 - \tilde{g} - \frac{1}{2}\tilde{n}\right) = 4 - 2\tilde{g} - \tilde{n}$ and $\tilde{n} \leqslant 3 - 2\tilde{g}$. If $\tilde{n} > 2$, then $\tilde{g} = 0$, $\tilde{n} = 3$

and $m = 1$. Therefore, the boundary of $V$ is connected and the closure of $V$ is homeomorphic to the disk $\Lambda$.                                          $\square$

THEOREM 4.2. *The dimension of any connected component of the space* $H^{\mathbb{R}}_{g,n}$ *is* $2(g + n - 1)$.

Proof. Let $H$ be a connected component of $H^{\mathbb{R}}_{g,n}$ and let $(P, \tau, f) \in H$. The critical values of $f$ form the set $s^0_1, \ldots, s^0_{k_0}, s^1_1, \ldots, s^1_k, s^2_1, \ldots, s^2_k$, where $\bar{s}^0_i = s^0_i$ and $\bar{s}^1_i = s^2_i$. Let $U^t_i \ni s^t_i$ be neighborhoods such that (1) $\overline{U}^0_i = U^0_i$, $\overline{U}^1_i = U^2_i$; (2) $U^{t_1}_{i_1} \cap U^{t_2}_{i_2} = \varnothing$ if $t_1 \neq t_2$ or $i_1 \neq i_2$; (3) $U^t_i \cap B_f = s^t_i$. Using Lemmas 4.2, 3.1, and 4.1, we construct an $n$-sheeted covering $f' \colon P \to S$ such that (1) $f'$ and $f$ coincide on the set $P \setminus f^{-1}\left(\bigcup_{t,i} U^t_i\right)$; (2) the covering $f'$ has only simple critical points; (3) $f'(\tau p) = \overline{f'(p)}$ for all $p \in P$. According to Theorem 4.1, the surface $P$ can be endowed with a complex structure $P'$ with respect to which $f'$ is a holomorphic and $\tau'$ is an antiholomorphic mapping. Choosing the neighborhoods $U^t_i$ sufficiently small we can obtain a real meromorphic function $(P', \tau', f') \in H^{\mathbb{R}}_{g,n}$ in an arbitrary given neighborhood $(P, \tau, f) \in W \subset H^{\mathbb{R}}_{g,n}$.

General functions in $W$ depend on $r_1$ real parameters and $r_2$ pairs of complex conjugate points, where $r_1 + 2r_2 = 2(g + n - 1)$ is the number of critical values of a general $n$-sheeted covering by a surface of genus $g$. Hence $\dim H = 2(g + n - 1)$.                                          $\square$

REMARK 4.1. Introduce an involution $\theta \colon H^{\mathbb{C}}_{g,n} \to H^{\mathbb{C}}_{g,n}$ on the space $H^{\mathbb{C}}_{g,n}$ by setting $\theta(P, f) = (\overline{P}, \tau_{\mathbb{R}}, f)$, where $\tau_{\mathbb{R}} \colon S \to S$ is the involution of complex conjugation, and $\overline{P}$ is the Riemann surface obtained from $P$ by replacing each holomorphic local parameter $U \subset P$, $\phi \colon U \to \mathbb{C}$ by the complex conjugate parameter $\tau_{\mathbb{R}} \phi \colon U \to \mathbb{C}$. The involution $\theta$ is antiholomorphic with respect to the complex structure on $H^{\mathbb{C}}_{g,n}$ (cf. Remark 3.1). Hence the pair $(H^{\mathbb{C}}_{g,n}, \theta)$ can be treated as a "real space". The real points of this space, that is, the fixed points of the involution $\theta$, form the space $H^{\mathbb{R}}_{g,n}$.

3.     Recall that the topological type of a real meromorphic function $(P, \tau, f) \in H^{\mathbb{R}}_{g,n}$ was defined in the Introduction as a set of numbers $(g, n, \varepsilon \mid i_1, \ldots, i_k)$, where $(g, k, \varepsilon)$ is the topological type of the curve $(P, \tau)$ and $i_j$ is the index of the function $f$ on the oval $a_j \subset P^\tau$ of $(P, \tau)$, that is, the number of rotations (taken with sign, if $\varepsilon = 1$) of the point $f(p)$ along the contour $S_0 = \mathbb{R} \cup \infty$ under a single passage of $p$ along $a$.

We denote by $H(g, n, \varepsilon \mid i_1, \ldots, i_k)$ the space of all real meromorphic functions of topological type $(g, n, \varepsilon \mid i_1, \ldots, i_k)$.

Obviously, a connected component of $H^{\mathbb{R}}_{g,n}$ consists of meromorphic functions of the same topological type.

A meromorphic function $(P, \tau, f) \in H^{\mathbb{R}}_{g,n}$ belongs to the set $H^{\sharp}_{g,n}$ if (1) the covering $f$ has only simple critical points; (2) there are no critical points on ovals of positive index; (3) there are exactly two critical points on ovals of zero index; (4) the image $f(a)$ of any oval of zero index is a segment.

A meromorphic function $(P, \tau, f) \in H_{g,n}^{\sharp}$ belongs to the set $H_{g,n}^{*}$ if the images $f(a)$ and $f(b)$ of any two ovals $a$ and $b$ of zero index do not intersect each other: $f(a) \cap f(b) = \varnothing$.

For $(P, \tau, f) \in H_{g,n}^{\mathbb{R}}$ we set $Q(P, \tau, f) = \max_{s \in \mathbb{R} \cup \infty}(F_1(s) + F_2(s)) + F_3$, where $F_1(s)$ is the cardinality of the set $f^{-1}(s) \cap P^{\tau}$; $F_2(s) = -2r$, where $r$ is the number of ovals of zero index containing points of the set $f^{-1}(s)$; $F_3$ is the number of degenerate critical points of the covering $f$.

LEMMA 4.4. *Let $H$ be a connected component of the space $H_{g,n}^{\mathbb{R}}$. Then $H \cap H_{g,n}^{*} \neq \varnothing$.*

Proof. Let us prove first that $H \cap H_{g,n}^{\sharp} \neq \varnothing$. Let $(g, n, \varepsilon | i_1, \ldots, i_k)$ be a type of a meromorphic function in $H$. Then for $(P, \tau, f) \in H$ the inequality $Q(P, \tau, f) \geqslant \sum_{j=1}^{k} |i_j|$ holds and $(P, \tau, f) \in H \cap H_{g,n}^{\sharp}$ if and only if $Q(P, \tau, f) = \sum_{j=1}^{k} |i_j|$. Let $(P, \tau, f) \in H$ and let $Q(P, \tau, f) > \sum_{j=1}^{k} |i_j|$. Then at least one of the following situations realizes: (1) the covering $f$ has a degenerate critical point; (2) an oval of positive index contains a simple critical point; (3) an oval of zero index contains three simple critical points; (4) there is an oval $a$ of zero index such that $f(a) = S_0$. Let us prove that in each of these cases there is a function $(P', \tau', f') \in H$ such that $Q(P', \tau', f') \leqslant Q(P, \tau, f) - 1$.

In case (1) the meromorphic function $(P', \tau', f')$ can be constructed in the same way as in the proof of Theorem 4.2. If the covering $f$ has only simple critical points but satisfies one of the conditions (2)–(4), then the set $R_0 \subset S_0$, where the function $F_1(s) + F_2(s)$ reaches its maximum, decomposes into finitely many segments. Let $\ell$ be one of these segments. Consider its neighborhood $U$ whose closure is homeomorphic to the disk $\Lambda$ and such that $\overline{U} = U$ and $U \cap B_f \subset \ell$. The ends $s_1$ and $s_2$ of the segment $\ell$ are simple critical values. Let $V_i$ $(i = 1, 2)$ be the connected component of the preimage $f^{-1}(U)$ containing the point $p_i = f^{-1}(s_i) \cap A_f$. Let $V_1 = V_2$, and let $f|_{V_1}$ be a two-sheeted covering. Then $a = f^{-1}(\ell) \cap V_1$ is an oval of zero index containing no critical points other than $p_1$ and $p_2$. Therefore, the segment $\ell$ cannot be a connected component of the set of maximal values of the function $F_1(s) + F_2(s)$. If the covering $f|_{V_1}$ is not two-sheeted or $V_1 \neq V_2$, then, by Lemmas 4.2 and 4.3, the closure $V_1$ is homeomorphic to the disk $\Lambda$. Applying Lemma 4.1 and Theorem 4.1, one can construct a deformation $(P_t, \tau_t, f_t)$ of the function $(P, \tau, f)$ to a function $(P', \tau', f')$ such that $f_t|_{P \setminus V_1} = f|_{P \setminus V_1}$, where either $B_{f'} \cap S_0 = B_f \cap S_0 - (s_1 \cup s_2)$ (if $V_1 = V_2$), or $f'(V_1 \cap (P')^{\tau'}) \cap f'(V_2 \cap (P')^{\tau'}) \neq \varnothing$ (if $V_1 \neq V_2$). In both cases $Q(P', \tau', f') \leqslant Q(P, \tau, f) - 2$. Thus

$$\min_{H} Q(P, \tau, f) = \sum_{j=1}^{k} i_j, \quad H \cap H_{g,n}^{\sharp} \neq \varnothing.$$

Let $(P, \tau, f) \in H \cap H_{g,n}^{\sharp}$, let $a_1, \ldots, a_r$ be ovals of zero index of the function $(P, \tau, f)$, $s_i' \cup s_i'' = a_i \cap B_f$ $(i = 1, \ldots, r)$, and let $U_i' \supset s_i'$, $U_i'' \supset s_i''$ be neighborhoods such that (1) $\overline{U}_i' = U_i'$, $\overline{U}_i'' = U_i''$; (2) $U_i' \cap B_f = s_i'$, $U_i'' \cap B_f = s_i''$; (3) $U_i' \cup U_i'' \supset \ell_i = f(a_i)$; (4) $U_i' \cap U_j' = \varnothing$ for $i \neq j$. Denote by $V_i$ the connected component of the preimage $f^{-1}(U_i'')$ containing a critical point. Applying Lemmas 4.2 and 4.1 to the neighborhood $V_i$, we construct a continuous family of coverings $f_t^i \colon P \to S$ $(0 \leqslant t \leqslant 1)$ such that

$$f_0^i = f, \quad f_t^i|_{P \setminus V_i} = f^i|_{P \setminus V_i}, \quad f_t^i(\tau(p)) = \overline{f_t^i(p)}, \quad B_{f_1^i} \cap U_i'' \subset U_i'.$$

By Theorem 4.1, the homeomorphisms $f_t^i$ generate a continuous curve

$$\Phi_i \colon [0, 1] \to H_{g,n}^{\sharp}$$

such that

$$\Phi_i(0) = (P, \tau, f), \quad \Phi_i(1) = (P_i, \tau_i, f_i).$$

Then the ovals $a_1, \ldots, a_r$ are taken to ovals $a_j^i$ such that $f_i(a_j^i) = f(a_j)$ if $i \neq j$ and $f_i(a_i^i) \subset U_i^1$. Applying this argument successively for $i = 1, \ldots, r$, we obtain a function

$$(P', \tau', f') \in H \cap H_{g,n}^{\sharp},$$

whose ovals of zero index $a_1', \ldots, a_r'$ satisfy the condition $f'(a_i') \subset U_i'$, whence $(P', \tau', f') \in H \cap H_{g,n}^*$.          $\square$

LEMMA 4.5. *Let* $(P, \tau, f) \in H_{g,n}^*$, $s^1, \ldots, s^k \in B_f$ *and suppose*

$$\left( \bigcup_{m=1}^{k} s^m \right) \cap \left( \bigcup_{m=1}^{k} \overline{s^m} \right) = \varnothing.$$

*Let* $\phi^m \colon [0, 1] \to S$ $(m = 1, \ldots, k)$ *be continuous mappings such that* (1) $\phi^m(0) = s^m$; (2) $\phi^m([0, 1])$ *is a segment*; (3) *connected components* $\tilde{\ell}^m$ *of the preimage* $f^{-1}(\phi^m([0, 1]))$ *containing the point* $p^m = f^{-1}(s^m) \cap A_f$ *are pairwise disjoint and* $\tilde{\ell}^m \cap P^{\tau} = \varnothing$; (4) *for* $t \in (0, 1)$ *we have* $\phi^m(t) \in S - B_f$. *Then there is a continuous mapping*

$$\Phi \colon [0, 1] \to H_{g,n}^* \quad (\Phi(t) = (P_t, \tau_t, f_t))$$

*such that*

$$(P_0, \tau_0, f_0) = (P, \tau, f)$$

*and*

$$B_{f,t} = B_f - \bigcup_{m=1}^{k} (s^m \cup \overline{s^m}) + \bigcup_{m=1}^{k} (\phi^m(t) \cup \overline{\phi^m(t)}).$$

*If also* $\ell^m = \phi^m([0, 1])$ *is a cutting segment with respect to* $f$, *then for any* $t < 1$ *the segment* $\phi^m([t, 1])$ *is a cutting segment with respect to* $f_t$.

Proof. Consider a set of neighborhoods $U_1^m, \ldots, U_{r_m}^m$ covering together the segment $\ell^m = \phi^m([0, 1])$ and such that (1) $U_{2i-1}^m \cap (R \cup \infty) = \varnothing$, $\overline{U_{2i}^m} = U_{2i}^m$ $(i = 1, \ldots, (r_m - 1)/2)$; (2) $U_i^m \cap U_j^m \neq \varnothing$ if and only if $|i - j| = 1$; (3) $U_1^m \cap B_f = s^m$, $U_i^m \cap B_f = \varnothing$ for $1 < i < r_m$; (4) the sets $U_i^m \cap \ell^m$ are connected;

(5) the closure of $U_i^m$ is homeomorphic to the disk $\Lambda$; (6) $V_i^{m_1} \cap V_j^{m_2} = \varnothing$ for $m_1 \neq m_2$, where $V_i^m$ is a connected component of the preimage $f^{-1}(U_i^m)$, $m = 1, \ldots, r$, intersecting the set $\tilde{\ell}^m$. Let $t_i^m \in [0,1]$ be a point such that $s_i^m = \phi^m(t_i^m) \in U_i^m \cap U_{i+1}^m$.

Applying Lemma 3.1 to the sets $V_1^m$ and $\tau V_1^m$, we construct a continuous family of coverings $f_t \colon P \to S \, (0 \leqslant t \leqslant 1)$ such that

$$f_0 = f, \quad f_t \big|_{P \setminus \bigcup_{m=1}^k (V_1^m \cup \tau V_1^m)} = f \big|_{P \setminus \bigcup_{m=1}^k (V_1^m \cup \tau V_1^m)}, \quad f_t(\tau(p)) = \bar{f}_t(p)$$

for $p \in P$ and $B_{f_t} \cap U_1^m = \phi^m(t)$. By Theorem 4.1, the covering $f_t$ generates a continuous curve $\Phi \colon [0, t_1] \to H_{g,n}^* \ (\Phi(t) = (P_t, \tau_t, f_t))$ such that

$$B_{f_t} = B_f - \bigcup_{m=1}^k (s^m \cup \overline{s^m}) + \bigcup_{m=1}^t (\phi^m(t) \cup \overline{\phi^m(t)}),$$

and if $\phi^m([0,1])$ is a cutting segment with respect to $f$, then for all $t \in [0, t_1]$ the segment $\phi^m([t,1])$ is a cutting segment with respect to $f_t$.

Since $\tilde{\ell}^m \cap P^\tau = \varnothing$, we have $\tau_{t_1} V_2^m \neq V_2^m$ and, therefore, $\tau_{t_1} V_2^m \cap V_2^m = \varnothing$. Applying Lemma 3.1 to the neighborhoods $V_2^m$ and $\tau_{t_1} V_2^m$, we find a continuous family of homeomorphisms $f_t \colon P_{t_1} \to S \ (t_1 \leqslant t \leqslant t_2)$ such that

$$f_t \big|_{P_{t_1}} - \bigcup_{m=1}^k (V_2^m \cup \tau_{t_1} V_2^m) = f_{t_1} \big|_{P_{t_1}} - \bigcup_{m=1}^k (V_2^m \cup \tau_{t_1} V_2^m), \quad f_t(\tau_{t_1} p) = \overline{f_t(p)}$$

for $p \in P$ and $B_{f_t} \cap U_2^m = \phi^m(t)$. Applying Theorem 4.1, we find a continuous curve

$$\Phi \colon [t_1, t_2] \to H_{g,n}^* \quad (\Phi(t) = (P_t, \tau_t, f_t))$$

such that

$$B_{f_t} = B_f - \bigcup_{m=1}^k (s^m \cup \overline{s^m}) + \bigcup_{m=1}^k (\phi^m(t) \cup \overline{\phi^m(t)}),$$

and if $\phi^m([t_1, 1])$ is a cutting segment with respect to $f_{t_1}$, then for all $t \in [t_1, t_2]$ the segment $\phi^m([t, 1])$ is a cutting segment with respect to $f_t$.

Repeating this argument for the neighborhoods $U_3, U_4$ and so on, we obtain a continuous curve $\Phi \colon [0,1] \to H_{g,n}^*$ satisfying all the requirements of the lemma. $\qquad \square$

**4.** Let us describe six operations related to meromorphic functions.

(1) Let $f \colon P \to S$ be a meromorphic function on a compact Riemann surface $P$, let $a \subset P$ be a vanishing contour with respect to $f$, and let $f' \colon P' \to S$ be the reduction modulo this contour. Then the Riemann surface structure on $P$ defines a Riemann surface structure on $P'$ with respect to which $f'$ is a holomorphic mapping. Let $(P, \tau, f) \in H_{g,n}^*$ and let $\ell \subset S$ be a vanishing segment with respect to $f$ such that $\bar{\ell} = \ell$. The *reduction of a function* $(P, \tau, f)$ *modulo* $\ell$ is a real meromorphic function $(P', \tau', f')$, where $P'$ is the reduction of the surface $P$ modulo the contour $a_\ell$ of the segment $\ell$, and $\tau'$

and $f'$ are the mappings of the surface $P'$ induced by the mappings $\tau$ and $f$.

The *reduction of a function* $(P, \tau, f) \in H^*_{g,n}$ *modulo a set* $L = \{\ell_1, \ldots, \ell_m\}$ $\subset S$ of pairwise disjoint vanishing segments such that $\bar{L} = L$ is defined similarly.

(2) Let $f \colon P \to S$ be an $n$-sheeted covering and let $L = \{\ell_1, \ldots, \ell_k\} \subset P - A_f$ be a set of pairwise disjoint segments. Denote by $\tilde{f} \colon \tilde{P} \to S$ the $(n + k)$-sheeted covering obtained by extending the covering $f$ along the segments $\ell_1, \ldots, \ell_k$. If $(P, \tau, f)$ is a real meromorphic function and $\bar{L} = L$, then the involution $\tau$ composed with the complex conjugation involution on $S$ generates an involution $\tilde{\tau} \colon \tilde{P} \to \tilde{P}$ such that $\tilde{f}\tilde{\tau}(p) = \overline{\tilde{f}(p)}$ for all $p \in \tilde{P}$. By Theorem 4.1, the surface $\tilde{P}$ can be endowed with a complex structure making $(\tilde{P}, \tilde{\tau}, \tilde{f})$ into a real meromorphic function. We call this function the *extension of* $(P, \tau, f)$ *along the segments* $\ell_1, \ldots, \ell_k$.

(3) Set

$$S_+ = \{z \in \mathbb{C} \mid \operatorname{Im} z > 0\}, \quad S_- = \{z \in \mathbb{C} \mid \operatorname{Im} z < 0\}, \quad \bar{S}_\pm = S_\pm \cup S_0,$$
$$\mathbb{R}_+ = \{z \in \mathbb{R} \mid z > 0\}, \qquad \mathbb{R}_- = \{z \in \mathbb{R} \mid z < 0\}, \qquad \mathbb{R}_0 = \mathbb{R} \cup \infty.$$

Let $(P, \tau, f)$ be a real meromorphic function and let $\ell_1, \ell_2 \subset P - A_f$ be disjoint segments such that $f(\ell_1) = f(\ell_2) \subset S_0$ and $\tau \ell_1 = \ell_2$. Consider the compactification $\overline{P}$ of the complement $P - (\ell_1 \cup \ell_2)$ by the boundary contours $c_1$ and $c_2$. After identifying the points of the contours under the involution $\tau$ we obtain a meromorphic function $(P', \tau', f')$, which we call the *gluing of the function* $(P, \tau, f)$ *along the segments* $\ell_1$ *and* $\ell_2$.

The *gluing of the function* $(P, \tau, f)$ *along a union*

$$L = \{\ell_1^1, \ell_1^2, \ldots, \ell_m^1, \ell_m^2\} \subset P - A_f$$

of pairwise disjoint segments such that

$$f(\ell_i^1) = f(\ell_i^2) \subset S_0, \quad \tau \ell_i^1 = \ell_i^2$$

is defined similarly.

(4) Let $(P_1, f_1)$ and $(P_2, f_2)$ be complex meromorphic functions, and let $\phi \colon P_1 \to P_2$ be an invertible antiholomorphic mapping such that $f_2 \phi(p) = \overline{f_1(p)}$ for all $p$. Let $c_1, \ldots, c_r \subset P_1$ be segments pairwise intersecting each other and such that $f_1(\ell_j) \subset S_0$. Consider the compactification $\overline{P}_1$ of the complement $P_1 - \bigcup_{i=1}^r c_i$ by the contours $c_1^1, \ldots, c_r^1$ and the compactification $\overline{P}_2$ of the complement $P_2 - \bigcup_{i=1}^r \phi(c_i)$ by the contours $c_1^2, \ldots, c_r^2$. After identifying the surfaces $\overline{P}_1$ and $\overline{P}_2$ along the mapping we obtain a compact Riemann surface $P$, where $\phi$ induces an involution $\tau_1 \colon P \to P$ and the coverings $f_1$ and $f_2$ induce a covering $f \colon P \to S$. We say that the real meromorphic function $(P, \tau, f)$ is obtained from the functions $(P_1, f_1)$, $(P_2, f_2)$ and the mapping $\phi$ by *gluing along the segments* $c_1, \phi(c_1), \ldots, c_r, \phi(c_r)$.

(5) Let $P_+$ and $P_-$ be compact Riemann surfaces, maybe coinciding; let $\phi \colon P_+ \to P_-$ be an invertible antiholomorphic mapping, involutive if $P_+ =$

$P_-$, and let $f_\pm \colon P_\pm \to S$ be meromorphic functions such that $f_-\phi(p) = \overline{f_+(p)}$ for all $p \in P_+$, and suppose that if $P_+ = P_-$, then $f_+ = f_-$. Let $a_\pm \subset P_\pm$ (this relation, as well as other similar relations, must be understood as the pair of relations $a_+ \subset P_+$ and $a_- \subset P_-$) be one-sheeted segments with respect to $f_\pm$ such that $\phi(a_+) = a_-$ and $d_1, \ldots, d_k \subset S$ are segments not intersecting $f_\pm(a_\pm)$ and each other and such that $\overline{d_i} = d_i$, while the intersection $d_i \cap S_0$ consists of a single point. Set $b_\pm = f_\pm(a_\pm)$. Let $\tilde{P}_\pm$ be the compactification of the complement $P_\pm - a_\pm$ by the boundary contours $\tilde{a}_\pm$, and let $\tilde{S}$ be the compactification by the boundary contours $\tilde{b}_\pm$ of the complement $S - (b_+ \cup b_-)$. Extend the functions $f_\pm$ to continuous functions $\tilde{f}_\pm \colon P_\pm \to \tilde{S}$ and identify the contours $\tilde{a}_\pm$ with the contours $\tilde{b}_\pm$ by means of the functions $\tilde{f}_\pm|_{\tilde{a}_\pm}$. The mapping $\phi$ induces an antiholomorphic involution $\tau_0 \colon P_0 \to P_0$ on the resulting connected surface $P_0$. The covering $\tilde{f}_\pm$ and the identical coverings on $\tilde{S}$ induce a holomorphic covering $f_0 \colon P_0 \to S$. Let $(P, \tau, f)$ be a real meromorphic function obtained from the real meromorphic function $(P_0, \tau_0, f_0)$ by extending it along the segments $d_1, \ldots, d_k \subset S$. We say that the function $(P, \tau, f)$ is obtained from the functions $(P_+, f_+)$, $(P_-, f_-)$ and the mapping $\phi$ by *pasting an oval along the segments* $a_+, a_-, d_1, \ldots, d_k \subset S$. Under such a pasting, the real contour $S_0 \subset S$ is transformed into an oval of index $k + 1$ of the function $(P, \tau, f)$.

(6) Let $(P, \tau, f)$ be a real meromorphic function such that $B_f \cap S_0 = \varnothing$, let $P'$ be a connected component of the preimage $f^{-1}(\overline{S}_+)$, and let $a_1, \ldots, a_m$ be its boundary. Identify the contour $a_j$ with the boundary $S_0$ of the domain $S_-$ and extend the mapping $f|a_j$ from the boundary $S_0$ to a covering $f_j \colon \overline{S}_- \to \overline{S}_-$ so that $B_{f_i} \supset (-i)$. As a result, we obtain a covering $\tilde{f} \colon \tilde{P} \to S$ such that $B_{\tilde{f}} \cap S_- \supset (-i)$ and $\tilde{f}|_{P'} = f|_{P'}$, which we call the *standard extension of the function* $(P, \tau, f)$ *through the boundary of* $P'$.

THEOREM 4.3. *Let $H$ be a connected component of the space $H(g, n, \varepsilon | i_1,$ $\ldots, i_k)$, where $\prod_{j=1}^{k} i_j \neq \varnothing$, and either $\varepsilon = 0$ or $\sum_{j=1}^{k} |i_j| < n - 2$. Then there is a function $(P, \tau, f) \in H \cap H_{g,n}^*$ with the ovals $P^\tau = \{a_1, \ldots, a_k\}$ and pairwise disjoint segments $b_1, \ldots, b_k \subset S_+$ vanishing with respect to $f$ and such that the contours of the segments $b_j$ and $\overline{b}_j$ cut from the surface $P$ a domain $U_j$ homeomorphic to the sphere, where $U_j \cap f^{-1}(S_0) = a_j$, $j = 1, \ldots, k$.*

Proof. Let $(P, \tau, f) \in H \cap H_{g,n}^*$. We call the *defect* of the connected component $P_+$ of the preimage $f^{-1}(\overline{S}_+)$ the degree of $f$ on the part of the boundary $\varphi P_+ - P^\tau$. The *defect of the function* $(P, \tau, f)$ is the maximum of the defects of all connected components of the preimage $f^{-1}(\overline{S}_+)$ containing at least one oval of the curve. Let $(P, \tau, f)$ be a real meromorphic function where the maximum $\mu$ of the defect on the set $H \cap H_{g,n}^*$ is achieved. Let $P_+$ be a connected component of the preimage $f^{-1}(\overline{S}_+)$ containing the oval $a$ such that the degree of $f$ on the part $\varphi P_+ - P^\tau$ of the boundary is $\mu$.

Let us prove that $\mu \geqslant 2$. Indeed, if $\mu = 0$, then $P = P_+ \cup \tau P_+$, whence $\varepsilon = 1$ and $\sum_{j=1}^n |i_j| = n$. Suppose that $\mu = 1$. Then $b = \varphi P_+ - P^\tau$ is a closed contour one-sheeted with respect to $f$. Let $P_-$ be the connected component of the preimage $f^{-1}(\bar{S}_-)$ containing the contour $b$. If $\varphi P_- \cap P^\tau \neq \varnothing$, then the defect of the surface $\tau P_- \subset f^{-1}(\bar{S}_+)$ is 1 and $\varphi P_- - P^\tau = b$. Therefore, if $\varphi P_- \cap P^\tau \neq \varnothing$ or $\varphi P_- = b$, then $P = (P_+ \cup P_-) \cup \tau(P_+ \cup P_-)$, whence $f^{-1}(S_0) - P^\tau = b \cup \tau b$. In particular, $\varepsilon = 1$ and $\sum_{i=1}^k |i_j| = n - 2$, in contradiction with the assumptions of the theorem.

Thus $\varphi P_- \cap P^\tau = \varnothing$ and $\varphi P_- \neq b$. In particular, the restriction $f|_{P_-}$ is not one-sheeted and therefore $A_f \cap P_- \neq \varnothing$. Connect the point $s_0$ of the set $f(A_f \cap P_-)$ with the point $s_1 \subset S_+$ by a segment $\phi \colon [0,1] \to \ell \subset (S - B_f) \cup s_0$ intersecting the contour $S_0$ at a single point. The connected component $\tilde{\ell}$ of the preimage $f^{-1}(\ell)$ containing this point does not intersect the set $P^\tau$. Applying Lemma 4.5, we find a continuous path

$$\Phi \colon [0,1] \to H_{g,n}^{\mathbb{R}} \qquad (\Phi(t) = (P_t, \tau_t, f_t))$$

such that

$$(P_0, \tau_0, f_0) = (P, \tau, f) \quad \text{and} \quad B_{f_t} = B_f - (s_0 \cup \bar{s}_0) + (\phi(t) \cup \overline{\phi(t)}).$$

In the deformation process, the connected component $P_+$ is taken to the component $P_+^1 \subset f_1^{-1}(\bar{S}_+)$ whose boundary contains the same ovals as the component $P_+$. The complement $b^1 = \varphi P_+^1 - P_1^{\tau_1}$ also consists of a single contour, but the index of the function $f_1$ on this contour is equal to the sum of the indices of the function $f$ on the two connected components of the set $f^{-1}(S_0) \cap P_-$ intersected by the segment $\tilde{\ell}$. Hence, the defect of the surface $P_+^1$ is greater than 1, which contradicts the maximality assumption on the function $(P, \tau, f)$. Therefore, $\mu \geqslant 2$.

Consider the standard extension $f_+ \colon \tilde{P}_+ \to S$ of the function $(P, \tau, f)$ through the boundary of the domain $P_+$. Then, by Lemmas 2.1, 2.2, and Corollary 2.2, there is a vanishing segment $b$ with respect to $f_+$ cutting from the surface $\tilde{P}_+$ a domain $U$ homeomorphic to the sphere and such that $\tilde{U} \cap f_+^{-1}(S_0) = a$. Hence the contours of the vanishing segments $b$ and $\bar{b}$ with respect to $f$ cut from the surface $P$ a domain $U$ homeomorphic to the sphere and such that $U \cap P^\tau = a$.

This argument, together with the reduction modulo the segments $b$ and $\bar{b}$, can be considered to be the first and general induction step on $k$. $\qquad \square$

## 5. Connected components of spaces of real meromorphic functions

**1.** Let $(P, \tau, f)$ be a real meromorphic function on a separating curve. By definition, this means that the set $P - P^\tau$ decomposes into two connected components. Let $\tilde{P}_+$ be one of these components. The orientation of the contour $S_0$ given by increasing of real numbers induces an orientation of the sphere $S$. The covering $f$ pulls this orientation back to the domain $\tilde{P}_+$, which, in turn, induces an orientation of the boundary $\varphi \tilde{P}_+ = P^\tau$. The covering $f$ takes the contour $a \subset P^\tau$ oriented in this way to the oriented contour $S_0$. Recall that the *degree of the mapping $f$ on the oval $a$* (with respect to the domain $\tilde{P}_+$) is the number of complete rotations of the point $f(p)$ around the contour $S_0$ as the point $p$ makes a single path around the contour $a$ in the positive direction, taking the sign into account.

The set of numbers $(g, n, 1 | i_1, \ldots, i_k)$, where $(g, k, 1)$ is the topological type of the curve $(P, \tau)$, $n$ is the number of sheets of the covering $f$, and $i_1, \ldots, i_k$ is the unordered set of degrees of $f$ on the ovals of the curve $(P, \tau)$ (with respect to one of the connected components of the complement $P - P^\tau$) is called the *topological type of the function $(P, \tau, f)$*. A topological type $(g, n, 1 | i_1, \ldots, i_k)$ is well defined up to permutations and simultaneous change of sign of all the indices $i_j$. The topological type obviously is an invariant of a connected component of the space $H^{\mathbb{R}}_{g,n}$.

Denote by $H(g, n, 1 | i_1, \ldots, i_k)$ the space of all meromorphic functions of topological type $(g, n, 1 | i_1, \ldots, i_k)$.

THEOREM 5.1. *Let $(P, \tau, f) \in H(g, n, 1 | i_1, \ldots, i_k)$ and let*

$$\left| \sum_{j=1}^{k} i_j \right| < \sum_{j=1}^{k} |i_j| = n - 2.$$

*Let $P_+ = f^{-1}(\overline{S}_+) \cap \tilde{P}_+$, where $\tilde{P}_+$ is a connected component of the complement $P - P^\tau$. Then $P_+$ is a connected surface and its genus $\chi$ does not vary under continuous deformations of the function $(P, \tau, f)$.*

Proof. Assume that the degrees $i_j$ are defined with respect to the connected component $\tilde{P}_+$. The relation $\sum_{j=1}^{k} |i_j| = n - 2$ immediately implies that (1) $A_f \cap f^{-1}(S_0) \subset P^\tau$; (2) there are exactly two critical points on each oval of zero index and the projections of two such ovals do not intersect each other; (3) at most one oval of nonzero index contains critical points, and these points are either a single critical point of ramification order 3, or two critical points of ramification order 2.

Suppose that the ovals of nonzero index of the curve $(P, \tau)$ do not contain critical points. Consider the reduction $(P_0, \tau_0, f_0)$ of the function $(P, \tau, f)$ modulo all the ovals of zero index. Let $\tilde{P}^0_+$ be the image of $\tilde{P}_+$ under this reduction. Then $b_+ = f_0^{-1}(S_0) \cap \tilde{P}^0_+$ is a simple closed contour. It contains half of the traces of the reduction and separates the ovals where the degree of $f_0$ with respect to $\tilde{P}^0_+$ is positive and those where the degree is negative. Hence the contour $b_+$ cuts from the surface $\tilde{P}^0_+$ a connected surface $P_+ = f_0^{-1}(\overline{S}_+) \cap \tilde{P}_+ \neq \tilde{P}_+$.

By Lemma 4.4, to complete the proof it suffices to show that an arbitrary deformation

$$\Phi \colon [0, 1] \to H(g, n, 1 | i_1, \ldots, i_k) \quad (\Phi(t) = (P_t, \tau_t, f_t), \ \Phi(0) = (P, \tau, f))$$

preserves the connectedness and the genus of the surface $P_+$. These characteristics can vary only if the set $A_{f_t} \cap f^{-1}(S_0)$ varies. Hence we must consider the following three cases: (1) two ramification points of order 2 symmetric with respect to $\tau_t$ are taken to the same ramification point of order 3 on an oval of positive index; (2) symmetric ramification points are taken to a point of an oval of negative index; (3) a ramification point of order 3 decomposes into two simple ramification points belonging to an oval.

In the first case the contour

$$b_t = (f_t^{-1}(S_0) - P_t^{\tau_t}) \cap (P_t)_+$$

is tangent to an oval of positive degree. The connectedness of the set $(P_t)_+$ is preserved, the number of points in the set $A_{f_t} \cap (P_t)_+ \cap f_t^{-1}(S_+)$ decreases by one, as well as the number of connected components of the boundary $\varphi(P_t)_+$. It is easy to show, using the Riemann–Hurwitz formula, that the sum of the doubled genus of $(P_t)_+$ and the number of sheets of the covering $f_t|_{(P_t)_+}$ equals the difference between the number of points in the set $A_{f_t} \cap (P_t)_+ \cap f_t^{-1}(S_+)$ and the number of connected components of the boundary $\varphi(P_t)_+$. Hence in case (1) the genus of the surface $(P_t)_+$ remains the same. In the cases (2) and (3) neither the connectedness of the surface $(P_t)_+$, nor the number of points in the set $A_{f_t} \cap (P_t)_+ \cap f_t^{-1}(S_+)$, nor the number of connected components of the boundary $\varphi(P_t)_+$ vary.                    □

Let

$$\left| \sum_{j=1}^{k} i_j \right| < \sum_{j=1}^{k} |i_j| = n - 2.$$

The *extended topological type* of a function $(P,\tau,f) \in H(g,n,1|i_1,...,i_k)$ is the set of numbers $(g,n,1|i_1,...,i_k|\chi)$, where $\chi$ is the genus of the surface $P_+ = f^{-1}(\overline{S}_+) \cap \tilde{P}_+$ (where $\tilde{P}_+$ is the connected component of the complement $P - P^\tau$ (with respect to which the degrees $i_1,...,i_k$ are defined).

The definition of extended topological type allows for the same ambiguity as the definition of topological type; this ambiguity is related to the choice of the connected component $\tilde{P}_+$. Under the choice of the component, all the degrees change their sign and the genus $\chi$ is transformed into the genus $\frac{1}{2}(g - k + 1) - \chi$. Thus an extended topological type $(g,n,1|i_1,...,i_k|\chi)$ is well defined up to permutations of the numbers $i_j$ and the change

$$(g,n,1|i_1,...,i_k|\chi) \to \left( g,n,1| - i_1,...,-i_k| \frac{1}{2}(g - k + 1) - \chi \right).$$

By Theorem 5.1, the extended topological type is an invariant of a connected component of $H_{g,n}^{\mathbb{R}}$.

Denote by $H(g,n,1|i_1,...,i_k|\chi)$ the set of all real algebraic curves of extended topological type $(g,n,1|i_1,...,i_k|\chi)$.

**2.**   Let us give some examples of real meromorphic functions on separating curves.

(1) Let

$$(P_+, f_+) \in H^+_{g,n}(i_1, \ldots, i_k), \quad \text{where } n = \sum_{j=1}^{k} i_j$$

(see Remark 3.2). Replacing the complex structure on $P_+$ by the complex conjugate one and the mapping $f_+$ by the mapping $f_- = \tau_R f_+$ (where $\tau_R(z) = \bar{z}$), we obtain a complex meromorphic function $(P_-, f_-)$ and an invertible antiholomorphic mapping $\phi \colon P_+ \to P_-$ such that $f_- \phi(p) = \overline{f_+(P)}$ for all $p \in P_+$. Let us identify the boundary $\varphi(\tilde{P}_+)$ of the surface $\tilde{P}_+ = P_+ \cap f^{-1}(\bar{S}_+)$ with the boundary $\varphi(\tilde{P}_-)$ of the surface $\tilde{P}_- = P_- \cap f^{-1}(\bar{S}_-)$ under the mapping $\phi|_{\varphi P_+}$. Then we obtain a surface $P$ where $\phi$ induces an antiholomorphic involution $\tau \colon P \to P$, and the coverings $f_+$ and $f_-$ induce a covering $f \colon P \to S$. It is easy to see that the result is a real meromorphic function $(P, \tau, f) \in H(2g + n - 1, n, 1 | i_1, \ldots, i_k)$ and $\sum_{j=1}^{k} i_j = n$. Set

$$F_1(P_+, f_+) = (P, \tau, f).$$

(2) Let

$$(P', f') \in H^+_{g', n'}(1, i'_1, \ldots, i'_{k'}), \quad (P'', f'') \in H^+_{g'', n''}(1, i''_1, \ldots, i''_{k''}),$$

where

$$n' = \sum_{j=1}^{k'} i'_j + 1, \quad n'' = \sum_{j=1}^{k''} i''_j + 1, \quad k' + k'' > 0.$$

Let $c_1, \ldots, c_r \subset S_0$ be pairwise disjoint segments. Set

$$(P_1, \tau_1, f_1) = F_1(P', f') \quad \text{and} \quad (P_2, \tau_2, f_2) = F_1(P'', f'').$$

Consider two ovals of index one $a' \in P_1^{\tau_1}$ and $a'' \in P_2^{\tau_2}$. Let us compactify each of the complements $P_1 - a'$ and $P_2 - a''$ by two contours $a'_\pm$, $a''_\pm$ and extend the coverings $f_1$ and $f_2$ to these contours (the notation is chosen so that small neighborhoods of the contours $a'_+$ and $a''_+$ are taken by the mappings $f_1$ and $f_2$ to subsets of the domain $S_+$). Since the functions $f_1$ and $f_2$ are one-sheeted on the contours $a'_\pm$, $a''_\pm$, there are homeomorphisms $\phi' \colon a'_+ \to a''_-$ and $\phi'' \colon a''_+ \to a'_-$ such that

$$f_1|_{a'_+} = f_2 \phi' \quad \text{and} \quad f_2|_{a''_+} = f_1 \phi'.$$

Identifying the contours by means of these homeomorphisms, we obtain a surface $P_0$ where the involutions $\tau_1$ and $\tau_2$ induce an involution $\tau_0$, and the coverings $f_1$ and $f_2$ induce a covering $f_0$. It is easy to show that $(P_0, \tau_0, f_0)$ is a real meromorphic function and the set $f_0^{-1}(S_0) - P_0^{\tau_0}$ decomposes into two contours $a_1$ and $a_2$ one-sheeted with respect to $f_0$. Set

$$c_i^j = f_0^{-1}(c_i) \cap a_j$$

and denote by $F_2((P', f'), (P'', f''), c_1, \ldots, c_r)$ the gluing $(P, \tau, f)$ of the function $(P_0, \tau_0, f_0)$ along the segments $c_i^j$.

If $k''=0$, then $g''=0$ and

$$(P,\tau,f) \in H(2g' + k' + r - 1, n' + 1, 1 \,|\, i'_1,\dots,i'_k, \underbrace{0,\dots,0}_{r}),$$

and

$$\sum_{j=1}^{k'} i'_j = n' - 2.$$

If $k',k''>0$, then

$$\sum_{j=1}^{k'} i'_j + \sum_{j=1}^{k''} i''_j = n' + n'' - 2,$$

$$(P,\tau,f) \in H(2(g'+g'')+k'+k''-1, n'+n'', 1 \,|\, i'_1,\dots,i'_{k'}, -i''_1,\dots, -i''_{k''}, \underbrace{0,\dots,0}_{r} \,|\, g').$$

(3) Let $g_0 \geqslant 0$, $k_+ + k_- + r > 0$, $t \geqslant 0$, $i^+_j > 0$ $(j=1,\dots,k_+)$ and $i^-_j > 0$ $(j=1,\dots,k_-)$. Let $a_1,\dots,a_{g_0+1}, b^+_1,\dots,b^+_{k_+}, b^-_1,\dots,b^-_{k_-}, \ell_1,\dots,\ell_t, c_1,\dots,c_r \subset S$ be pairwise disjoint segments such that $b^{\pm}_j \subset S_{\pm}$ (this relation, as well as similar relations below, must be understood as two relations $b^+_j \subset S_+, b^-_j \subset S_-), c_j \subset S_0$. Let

$$d^{\pm}_{j,1},\dots,d^{\pm}_{j,i_j-1} \subset S$$

be pairwise disjoint segments not intersecting the segment $b^{\pm}_j$ and such that

$$\overline{d^{\pm}_{j,i}} = d^{\pm}_{j,i} \quad (j=1,\dots,k_{\pm}),$$

and the intersection $d^{\pm}_{j,i} \cap S_0$ consists of a single point.

Denote by $f_0 \colon P_0 \to S$ the two-sheeted (hyperelliptic) covering whose set of ramification points coincides with the set of the ends of the segments $a_1,\dots,a_{g_0+1}$. Then $(P_0,f_0) \in H^{\mathbb{C}}_{g,2}$. The preimage $f_0^{-1}\left(\bigcup_{i=1}^{g_0+1} a_i\right)$ decomposes the surface $P_0$ into two one-sheeted domains $U_1$ and $U_2$. Set

$$\tilde{b}^{\pm}_j = f_0^{-1}(b^{\pm}_j) \cap U_1, \quad \tilde{c}_j = f_0^{-1}(c_j) \cap U_1, \quad \text{and} \quad \tilde{\ell}_j = f_0^{-1}(\ell_j) \cap U_1.$$

Replacing the complex structure on $P_0$ by a complex conjugate one and the mapping $f_0$ by the mapping $f_1 = \tau_{\mathbb{R}} f_0$, we obtain a complex meromorphic function $(P_1,f_1) \in H^{\mathbb{C}}_{g,2}$ and an invertible antiholomorphic mapping $\phi \colon P_0 \to P_1$ such that $f_1 \phi(p) = \overline{f_0(p)}$ for all $p \in P_0$.

Denote by

$$D(a_1,\dots,a_{g_0+1} \,|\, b^+_1,\dots,b^+_{k_+}, |\, d^+_{1,1},\dots,d^+_{1,i^+_1-1}| \cdots |d^+_{k_+,1},\dots,d^+_{k_+,i^+_{k_+}-1} \,|\, b^-_1,\dots,$$

$$b^-_{k_-} \,|\, d^-_{1,1},\dots,d^-_{1,i^-_1-1}| \cdots |d^-_{k_-,1},\dots,d^-_{k_-,i^-_{k_-}-1} \,|\, c_1,\dots,c_r \,|\, \ell_1,\dots,\ell_t)$$

the real meromorphic function obtained from the complex meromorphic functions $(P_0,\tau_0)$ and $(P_1,\tau_1)$ and the antiholomorphic mapping $\phi$ by attaching the ovals along the segments

$$\tilde{b}^{\pm}_j, \phi(\tilde{b}^{\pm}_j), d^{\pm}_{j,1},\dots,d^{\pm}_{j,i^{\pm}_j-1} \quad (j=1,\dots,k_{\pm}),$$

gluing along the segments $\tilde{c}_j, \phi(\tilde{c}_j)$ and extending along the segments $\tilde{\ell}_j$ and $\phi(\tilde{\ell}_j)$.

Denote the set of all such functions by

$$D(g_0 \,|\, k_+ \,|\, i_1^+, \ldots, i_{k_+}^+ \,|\, k_- \,|\, i_1^-, \ldots, i_{k_-}^- \,|\, r \,|\, t).$$

By construction, this set is a connected subset in the space

$$H(2g+k_1+k_2+r-1, k_++k_-+2\ell+4, 1 \,|\, i_1^+, \ldots, i_{k_+}^+, -i_1^-, \ldots, -i_{k_-}^-, \underbrace{0, \ldots, 0}_{r}).$$

THEOREM 5.2. *A set of numbers* $(g, n, 1 \,|\, i_1, \ldots, i_k)$ *is the topological type of a real algebraic function if and only if one of the following cases realizes:*

(1) $n = 1$, $g = 0$, $k = i_1 = 1$;

(2) $n = 2$, $k = g + 1$, $i_1 = \cdots = i_k = 0$;

(3) $n \geqslant 2$, $1 \leqslant g + 1$, $k \equiv g + 1 \pmod 2$,

$$\left| \sum_{j=1}^{k} i_j \right| = \sum_{j=1}^{k} |i_j| = n, \quad \prod_{j=1}^{k} i_j \neq 0;$$

(4) $n \geqslant 3$, $1 \leqslant k \leqslant g + 1$, $k \equiv g + 1 \pmod 2$,

$$\sum_{j=1}^{k} |i_j| \leqslant n - 2, \quad \sum_{j=1}^{k} i_j \equiv n \pmod 2.$$

*A set of numbers* $(g, n, 1 \,|\, i_1, \ldots, i_k \,|\, \chi)$ *is the extended topological type of a real meromorphic function if and only if*

$$1 \leqslant k \leqslant g + 1, \quad k \equiv g + 1 \pmod 2,$$

$$\left| \sum_{j=1}^{k} i_j \right| < \sum_{j=1}^{k} |i_j| = n - 2, \quad 0 \leqslant \chi \leqslant \frac{1}{2}(g - k + 1).$$

Proof. The relations $1 \leqslant k \leqslant g + 1$ and $k \equiv g + 1 \pmod 2$ hold for all separating curves (cf. Section 1 of Chapter 1). For $n = 1$ the function $f$ is the identity function. For $n = 2$ the classification of real meromorphic functions coincides with the classification of real hyperelliptic curves and is well known, see e.g. [58]. Let $n \geqslant 3$ and let $(P_0, \tau_0, f_0)$ be a meromorphic function of topological type $(g, n, 1 \,|\, i_1, \ldots, i_k)$. By Lemma 4.4, there is a function $(P, \tau, f) \in H_{g,n}^*$ of the same topological type. The preimage $f^{-1}(s)$ of a point $s \in R - B_f$ contains $n$ points and consists of (1) points belonging to ovals of positive index; (2) points belonging to ovals of index zero; (3) points not belonging to ovals. The number of points of the first type is $\sum_{j=1}^{k} i_j$. The points of the second type decompose into pairs of points belonging to the same oval. Points of the third type decompose into pairs of points transposed by the involution $\tau$. Hence

$$\sum_{j=1}^{k} |i_j| \leqslant n \quad \text{and} \quad \sum_{j=1}^{k} i_j \equiv n \pmod 2.$$

Suppose $\sum_{j=1}^{k}|i_j|=n$. Then $P^\tau=f^{-1}(S_0)$, and therefore, $\prod_{j=1}^{k}i_j\neq 0$, and $f^{-1}(S_+)$ is a connected component of the complement $P-P^\tau$. Thus the covering $f_k|_{P^\tau}\colon P^\tau\to S_0$ preserves the orientation of $P^\tau$ given by the surface $f^{-1}(S_+)$, i.e.,

$$\left|\sum_{j=1}^{k}i_j\right|=\sum_{j=1}^{k}|i_j|.$$

If $(P,\tau,f)\in H(g,n,1|i_1,\ldots,i_k|\chi)$, then $\chi$ is the genus of a part of the connected component of the complement $P-P^\tau$, whose genus equals $\frac{1}{2}(g-k+1)$. Therefore, $0\leqslant\chi\leqslant\frac{1}{2}(g-k+1)$.

The constructions in the beginning of this section produce examples of curves of an arbitrary topological type and extended topological type satisfying the assumptions of the theorem. □

**3.** Let us prove that the topological characteristics just introduced describe connected components of the space of real meromorphic functions on separating real algebraic curves. In this section we consider only sets of numbers $(g,n,1|i_1,\ldots,i_k)$ and $(g,n,1|i_1,\ldots,i_k|\chi)$ satisfying the assumptions of Theorem 5.2.

For $n=2$ the space $H(g,n,1|i_1,\ldots,i_k)$ consists of real hyperelliptic functions with constant number of critical values. The connectedness of this space is obvious.

THEOREM 5.3. *The space* $H(g,n,1|i_1,\ldots,i_k)$ *is connected if* $\sum_{j=1}^{k}|i_j|=n$.

Proof. Suppose $(P,\tau,f)\in H(g,n,1|i_1,\ldots,i_k)$, where $\sum_{j=1}^{k}|i_j|=n$. Then $f^{-1}(S_+)$ is a connected component of the complement $P-P^\tau$ and $(P,\tau,f)=F_1(P_+,f_+)$, where $(P_+,f_+)$ is the extension of the function $(P,\tau,f)$ through the boundary of the domain $f^{-1}(\bar S_+)$. Hence

$$H(g,n,1|i_1,\ldots,i_k)=F_1(H^+_{(1/2)(g-k+1),n}(i_1,\ldots,i_k)).$$

Since the space

$$H^+_{(1/2)(g-k+1),n}(i_1,\ldots,i_k)$$

is connected (cf. Remark 3.2), the last equation implies that also the space $H(g,n,1|i_1,\ldots,i_k)$ is connected. □

THEOREM 5.4. *The space* $H(g,n,1|i_1,\ldots,i_k)$ *is connected for* $\left|\sum_{j=1}^{k}i_j\right|=\sum_{j=1}^{k}|i_j|=n-2$. *The space* $H(g,n,1|i_1,\ldots,i_k|\chi)$ *is connected.*

Proof. Without loss of generality one can assume that

$$(i_1,\ldots,i_k)=(i'_1,\ldots,i'_{k'},-i''_1,\ldots,-i''_{k''},\underbrace{0,\ldots,0}_{r}),$$

where $i'_j,i''_j>0$ and $k'\geqslant k''$. Set

$$n'=\sum_{j=1}^{k'}i'_j+1,\quad n''=\sum_{j=1}^{k''}i''_j+1.$$

Let $H$ be a connected component either of the space $H(g,n,1|i_1,\ldots,i_k|\chi)$ or of the space $H(g,n,1|i_1,\ldots,i_k)$, where $\sum_{j=1}^{k}|i_j|=n-2$. By Lemma 4.4, $H\cap H^*_{g,n}\neq\varnothing$. Let $(P,\tau,f)\in H\cap H^*_{g,n}$, let $\tilde{c}_1,\ldots,\tilde{c}_r$ be all ovals of zero index of the function $(P,\tau,f)$, $c_j=f(\tilde{c}_j)$, and let $(P_0,\tau_0,f_0)$ be the reduction of $(P,\tau,f)$ modulo these ovals. The set $f_0^{-1}(\overline{S}_+)$ decomposes into two connected components $\tilde{P}'$ and $\tilde{P}''$ (Theorem 5.1). Without loss of generality we can assume that the ovals with indices $i'_1,\ldots,i'_{k'}$ belong to the boundary of the domain $\tilde{P}'$. Denote by

$$(P',f')\in H^+_{g',n'}(1,i'_1,\ldots,i'_{k'}),\quad (P'',f'')\in H^+_{g'',n''}(1,i''_1,\ldots,i''_{k''})$$

the extensions of the functions $(P_0,\tau_0,f_0)$ through the boundaries of the domains $\tilde{P}'$ and $\tilde{P}''$, respectively. Then, by definition,

$$(P,\tau,f)=F_2((P',f'),(P'',f''),c_1,\ldots,c_r).$$

Hence

$$H\cap H^*_{g,n}\subset F_2(H^+_{g',n'}(1,i'_1,\ldots,i'_{k'}),H^+_{g'',n''}(1,i''_1,\ldots,i''_{k''}),I^r),$$

where $I^r$ is the set of all sets of disjoint segments $c_1,\ldots,c_r\subset S_0$;

$$g'=\begin{cases}\frac{1}{2}(g-k+1) & \text{if }|\sum_{j=1}^{k}i_j|=\sum_{j=1}^{k}|i_j|,\\ \chi & \text{if }|\sum_{j=1}^{k}i_j|<\sum_{j=1}^{k}|i_j|,\end{cases}$$

and

$$g''=\frac{1}{2}(g-k+1)-g'.$$

According to Remark 3.2, this set is connected.

In the case

$$\left|\sum_{j=1}^{k}i_j\right|=\sum_{j=1}^{k}|i_j|$$

this implies that each connected component of the space $H(g,n,1|i_1,\ldots,i_k)$ contains the same connected set

$$F_2(H^+_{(1/2)(g-k+1),n'}(1,i'_1,\ldots,i'_{k'}),H^+_{0,n''}(1,i''_1,\ldots,i''_{k''}),I^r).$$

Therefore, the set $H(g,n,1|i_1,\ldots,i_k)$ is connected.

In the case

$$\left|\sum_{j=1}^{k}i_j\right|<\sum_{j=1}^{k}|i_j|$$

each connected component of the space $H(g,n,1|i_1,\ldots,i_k|\chi)$ contains the same connected set

$$F_2(H^+_{\chi,n'}(1,i'_1,\ldots,i'_{k'}),H^+_{(1/2)(g-k+1)-\chi,n''}(1,i''_1,\ldots,i''_{k''}),I^r)$$

and, therefore, also the space $H(g,n,1|i_1,\ldots,i_k|\chi)$ is connected. $\square$

THEOREM 5.5. *The space $H(g,n,1|i_1,\ldots,i_k)$ is connected for $\sum_{j=1}^{k}|i_j| \leqslant n - 4$.*

Proof. Without loss of generality we can assume that

$$(i_1,\ldots,i_k) = (i_1^+,\ldots,i_{k^+}^+,-i_1^-,\ldots,-i_{k^-}^-,\underbrace{0,\ldots,0}_{r}),$$

where $i_j^\pm > 0$. Let $H$ be a connected component of $H(g,n,1|i_1,\ldots,i_k)$, and let $(P,\tau,f) \in H \cap H_{g,n}^*$ (cf. Lemma 4.4). Let $c_1^0,\ldots,c_r^0$ be all ovals of zero index of the function $(P,\tau,f)$, $c_j = f(c_j^0)$, and let $(P_0,\tau_0,f_0)$ be the reduction of $(P,\tau,f)$ modulo these ovals.

By Theorem 4.3, there is a continuous path

$$\Phi\colon [0,1] \to H(g,n,1|i_1^+,\ldots,i_{k^+}^+,-i_1^-,\ldots,-i_{k^-}^-) \quad (\Phi(t) = (P_t,\tau_t,f_t))$$

such that $(P_1,\tau_1,f_1) \in H_{g-r,n}^*$, and there are segments $b_1^\pm,\ldots,b_{k_\pm}^\pm \subset S_\pm$ vanishing with respect to $f_1$ and such that the contours of the segments $b_j^\pm$ and $\overline{b_j^\pm}$ cut from the surface $P_1$ a domain $U_j^\pm$ homeomorphic to the sphere

$$U_j^\pm \cap f_1^{-1}(S_0) = a_j^\pm,$$

where $a_j^\pm$ is an oval of index $i_j^\pm$. (Hereinafter, these relations must be understood as a pair of relations: one for the sign "+", and the other for the sign "$-$", $j = 1,\ldots,k_\pm$.) Since $S_0 \neq \bigcup_{j=1}^{r} c_j$, the path $\Phi$ can be chosen in such a way that $B_{f_t} \cap c_j = \varnothing$ for all $t$. In this case, by Lemma 4.5, the path $\Phi$ induces a continuous path

$$\tilde{\Phi}\colon [0,1] \to H(g,n,1|i_1,\ldots,i_k)$$

with the beginning $(P,\tau,f)$ and the end $(\tilde{P},\tilde{\tau},\tilde{f}) \in H \cap H_{g,n}^*$.

Consider the set $f_1(A_{f_1} \cap U_j^\pm)$ and connect its points in pairs by disjoint segments

$$d_{j,1}^\pm,\ldots,d_{j,i_j^\pm-1}^\pm$$

such that $\overline{d_{j,m}^\pm} = d_{j,m}^\pm$. Denote by

$$(P',f'),(P'',f'') \in H_{\frac{1}{2}(g-k+1),\frac{1}{2}\left(n-\sum_{j=1}^{k}i_j\right)}^0$$

the reductions of the function $(P_1,\tau_1,f_1)$ modulo the segments of the form $b_j^\pm,\overline{b_j^\pm}$, not coinciding with $U_j^\pm$. Let $\tilde{c}_j$ and $\tilde{b}_j^\pm$ be the traces of the reduction modulo the segments $c_j$ and $b_j^\pm$ belonging to the surface $P'$. By Lemma 1.6, there is a set of segments $a_1,\ldots,a_{g_0+1}$ vanishing with respect to $f'$ ($g_0 = \frac{1}{2}(g-k+1)$) and cutting segments $\ell_1,\ldots,\ell_t$ with respect to $f'$ ($t = \frac{1}{2}(n - \sum_{j=1}^{k}|i_j|)$) so that the preimage

$$(f')^{-1}\left(\bigcup_{i=1}^{g_0+1} a_i\right)$$

separates from the surface $P'$ a domain that contains no segments of the form $\tilde{c}_j$ and $\tilde{b}_j^{\pm}$ and no films of the segments $\ell_i$, and the films of the segments $\ell_i$ also contain no segments of the form $\tilde{c}_j$ and $\tilde{b}_j^{\pm}$. But then

$$(\tilde{P}, \tilde{\tau}, \tilde{f}) = D(a_1, \ldots, a_{g_0+1} | b_1^+, \ldots, b_{k_+}^+ | d_{1,1}^+, \ldots, d_{1,i_1^+-1}^+ | \cdots | d_{k_+,1}^+, \ldots,$$
$$d_{k_+,i_{k_+}^+-1}^+ | b_1^-, \ldots, b_{k_-}^- | d_{1,1}^-, \ldots, d_{1,i_1^--1}^- | \cdots | d_{k_-,1}^-, \ldots,$$
$$d_{k_-,i_{k_-}^--1}^- | c_1, \ldots, c_r | \ell_1, \ldots, \ell_t).$$

Therefore,

$$H \cap D(g_0 | k_+ | i_1^+, \ldots, i_{k_+}^+ | k_- | i_1^-, \ldots, i_{k_-}^- | r | t) \neq \varnothing.$$

Hence all connected components of the set $H(g, n, 1 | i_1, \ldots, i_k)$ intersect the same connected component. Therefore, the set $H(g, n, 1 | i_1, \ldots, i_k)$ is connected. $\square$

**4.** Now let us consider real meromorphic functions $(P, \tau, f)$ on real algebraic nonseparating curves. By definition, this means that the set $P - P^\tau$ is connected. The *topological type of a function* $(P, \tau, f)$ is a set of numbers $(g, n, 0 | i_1, \ldots, i_k)$, where $(g, k, 0)$ is the topological type of the curve $(P, \tau)$, $n$ is the number of sheets of the covering $f$, and $i_1, \ldots, i_k \geq 0$ are the indices of $f$ on the ovals $a_j \subset P^\tau$, i.e., the number of paths of the point $f(p)$ along the contour $S_0$ as the point $p$ passes once along the contour $a$. The topological type $(g, n, 0 | i_1, \ldots, i_k)$ is defined up to a permutation of the indices $i_1, \ldots, i_k$.

We give examples of real meromorphic functions on nonseparating real algebraic curves.

Let

$$g_0 \geqslant 0, \quad k \geqslant 0, \quad i_j > 0 \ (j = 1, \ldots, k), \quad r \geqslant 0, \quad t \geqslant 0.$$

Also let

$$a_1, \ldots, a_{g_0+1}, b_1, \ldots, b_k, c_1, \ldots, c_r, \ell_1, \ldots, \ l_t \subset S$$

be pairwise disjoint segments, and suppose that for all $j, b_j, \ell_j \subset S_+$ we have $c_j \subset S_0$ and at most one segment $a_j$ has common points with the contour $S_0$, $\bigcup_{i=1}^{g_0+1} \bar{a}_i = \bigcup_{i=1}^{g_0+1} a_i$. Let $d_{j,1}, \ldots, d_{j,i_j-1} \subset S$ be pairwise disjoint segments not intersecting the segment $b_j$, and let $\bar{d}_{j,m} = d_{j,m}$.

Denote by $f_0 \colon P_0 \to S$ the two-sheeted (hyperelliptic) covering whose set of ramification points coincides with the set of ends of the segments $a_1, \ldots, a_{g_0+1}$. The preimage $f_0^{-1}\left(\bigcup_{i=1}^{g_0+1} a_i\right)$ cuts the surface $P_0$ into two domains $U_1$ and $U_2$ one-sheeted with respect to $f_0$. Set

$$\tilde{b}_j = f_0^{-1}(b_j) \cap U_1, \quad \tilde{c}_j = f_0^{-1}(c_j) \cap U_1, \quad \text{and} \quad \tilde{\ell}_j = f_0^{-1}(\ell_j) \cap U_1.$$

The complex conjugation involution induces involutions on the domains $U_1$ and $U_2$ hence defining an involution $\tau_0 \colon P_0 \to P_0$ such that $(P_0, \tau_0, f_0) \in H(g, 2, 0 | 0)$.

Denote by

$$D_0(a_1,\ldots,a_{g_0+1}|b_1,\ldots,b_k|d_{1,1},\ldots,d_{1,i_1-1}|\cdots|d_{k,1},\ldots,d_{k,i_k-1}|c_1,\ldots,c_r|\ell_1,\ldots,\ell_t)$$

the real meromorphic function obtained from $(P_0,\tau_0,f_0)$ by attaching the
ovals along the segments $\tilde{b}_j$, $\tau_0\tilde{b}_j$, $d_{j,1}$, $\ldots$, $d_{j,i_j-1}$ $(j=1,\ldots,k)$, gluing along
the segments $\tilde{c}_j$, $\tau_0\tilde{c}_j$ $(j=1,\ldots,r)$ and extending along the segments $\tilde{\ell}_j$, $\tau_0\tilde{\ell}_j$
$(j=1,\ldots,t)$. Denote the set of all such functions by $D_0(g_0|k|i_1,\ldots,i_k|r|t)$.
By construction, this set is a connected subset of the set

$$H(g_0 + k + r, k + r + 2\ell + 2, 0 | i_1,\ldots,i_k, \underbrace{0,\ldots,0}_{r}).$$

THEOREM 5.6. *A set of numbers* $(g,n,0|i_1,\ldots,i_k)$ *is the topological type
of a real meromorphic function if and only if* $0 \leqslant k \leqslant g$, $\sum_{j=1}^{k} i_j \leqslant n - 2$, *and*
$\sum_{j=1}^{k} i_j \equiv n \pmod 2$.

Proof. The necessity of the relations is proved in the same way as in the
case of functions on separating curves (cf. Theorem 5.2). Curves of the form
$D_0(a_1,\ldots,\ell_t)$ give examples of curves of arbitrary topological type satisfying
the assumptions of the theorem.                                              □

THEOREM 5.7. *The space* $H(g,n,0|i_1,\ldots,i_k)$ *is connected.*

Proof. Without loss of generality we can assume that $i_j > 0$ for $j \leqslant
k_+$ and $i_j = 0$ for $j > k_+$. Let $H$ be a connected component of the space
$H(g,n,0|i_1,\ldots,i_k)$ and let $(P,\tau,f) \in H \cap H_{g,n}^*$ (cf. Lemma 4.4). Let $c_1^0,\ldots,$
$c_{k-k_+}^0$ be all the ovals of zero index of the function $(P,\tau,f)$, let $c_j = f(c_j^0)$, let
$(P_0,\tau_0,f_0)$ be the reductions of $(P,\tau,f)$ modulo these ovals, and let $\tilde{c}_j,\tau_0\tilde{c}_j$
be the traces of the reduction. By Theorem 4.3, there is a continuous path

$$\Phi\colon [0,1] \to H(g,n,0|i_1,\ldots,i_k) \quad (\Phi(t) = (P_t,\tau_t,f_t))$$

such that $(P_1,\tau_1,f_1) \in H \cap H_{g-r,n}^*$, and there are pairwise disjoint segments
$b_1,\ldots,b_{k_+} \subset S_+$ vanishing with respect to $f_1$ and possessing the following
properties: the contours of the segments $b_j$ and $\tilde{b}_j$ cut from the surface $P_1$
a domain $U_j$ homeomorphic to the sphere and such that $U_j \cap f_1^{-1}(S_0)$ is an
oval of index $i_j$.

Since $S_0 \neq \bigcup_{j=1}^{k-k_+} \tilde{c}_j$, the path $\Phi$ can be chosen in such a way that $B_{f_t} \cap
c_j = \varnothing$ for all $t$. By Lemma 4.5, such a path induces a continuous path
$\tilde{\Phi}\colon [0,1] \to H(g,k,0|i_1,\ldots,i_k)$ with the beginning $(P,\tau,f)$ and end $(\tilde{P},\tilde{\tau},\tilde{f}) \in
H_{g,n}^* \cap H$.

Consider the set $f_1(A_{f_1} \cap U_1)$ and connect its points in pairs by continuous segments $d_{j,1},\ldots,d_{j,i_j-1}$ such that $\bar{d}_{j,m} = d_{j,m}$.

Let $(P',\tau',f') \in H\left(g - k, n - \sum_{j=1}^{k} i_j, 0|\right)$ be a reduction (not coinciding
with $U_j$) of the function $(P_1,\tau_1,f_1)$ modulo the segments $b_j, \bar{b}_j$ and $\tilde{b}_j$, and
let $\tau'\tilde{b}_j$ be the traces of this reduction. According to Lemma 2.5, there are

pairwise disjoint segments

$$\ell'_1, \ldots, \ell'_m \subset S \quad \left( m = \frac{1}{2} \left( n - \sum_{j=1}^{k} i_j - 2 \right) \right)$$

cutting with respect to $f'$ and such that the set $V'$ of the ends of these segments satisfies the conditions $\overline{V}' \cap V' = \varnothing$ and the films of these segments do not contain the traces of the reduction $\tilde{c}_j, \tau_0 \tilde{c}_j, \tilde{b}_j, \tau' \tilde{b}_j$. By Lemma 4.5, there is a continuous path $\Phi \colon [0,1] \to H(g-k, 2t+2, 0|)$ starting at $(P', \tau', f')$ and ending at $(P'', \tau'', f'')$ and taking the segments $\ell'_1, \ldots, \ell'_m$ to segments $\ell_1, \ldots, \ell_m \subset S_+$ cutting with respect to $f''$. Since $S \neq \bigcup_{j=1}^{k-k_+} c_j$, this path can be chosen so that for all $t$, $B_{f_t} \cap c_j = B_{f_t} \cap B_{f_t} \cap \bar{b}_j = \varnothing$. According to Lemma 4.5, such a path induces a continuous path $\tilde{\Phi}_0 \colon [0,1] \to H(g, n, 0|i_1, \ldots, i_k)$ with the beginning $(\tilde{P}, \tilde{\tau}, \tilde{f})$ and the end $(\tilde{P}_1, \tilde{\tau}_1, \tilde{f}_1)$.

Let $(P_2, \tau_2, f_2) \in H(g-k, 2, 0|)$ be the reduction of $(P'', \tau'', f'')$ modulo the segments $\ell_j, \bar{\ell}_j, j = 1, \ldots, m$. By Lemma 2.5, there is a set $a_1, \ldots, a_{g-k+1}$ of pairwise disjoint segments, vanishing with respect to $f_2$, at most one of which intersects the contour

$$S_0, \quad \bigcup_{i=1}^{g-k+1} \overline{a_i} = \bigcup_{i=1}^{g-k+1} a_i,$$

and the contours of the segment $a_i$ cut the domain $P_2$ into domains $U_1$ and $U_2$, both one-sheeted with respect to $f_2$, the first of which, $U_1$, contains the traces of the reductions modulo all segments of the form $b_j, c_j, \ell_j, \bar{b}_j, \bar{\ell}_j$. But then

$$(\tilde{P}_1, \tilde{\tau}_1, \tilde{f}_1) = D_0(a_1, \ldots, a_{g-k+1} | b_1, \ldots, b_{k_+} | d_{1,1}, \ldots, d_{1,i_1-1} | \cdots$$
$$| d_{k_+,1}, \ldots, d_{k_+,i_{k_+}-1} | c_1, \ldots, c_{k-k_+} | \ell_1, \ldots, \ell_m),$$

and therefore,

$$H \cap D_0(g-k|k_+|i_1, \ldots, i_{k_+}|k-k_+|m) \neq \varnothing.$$

Hence each connected component of the space $H(g, n, 0|i_1, \ldots, i_k)$ intersects the same connected set $D_0(g-k|k_+|i_1, \ldots, i_{k_+}|k-k_+|m)$. Therefore, the space $H(g, n, 0| i_1, \ldots, i_k)$ is connected. $\qquad\square$

**5.** Besides the involution of complex conjugation $\tau_{\mathbb{R}}(s) = \bar{s}$, there is another involution on the Riemann sphere $S = \mathbb{C} \cup \infty$ topologically distinct from the first, namely, the antiholomorphic involution $\tau_*(s) = -\bar{s}^{-1}$. Define, by analogy with real meromorphic functions, a *pseudoreal meromorphic function* on a real algebraic curve as a triple $(P, \tau, f)$, where $(P, \tau)$ is a real algebraic curve and $f \colon P \to S$ is a function such that $f\tau = \tau_* f$.

Similarly to the involution of complex conjugation $\tau_{\mathbb{R}}$ (cf. Remark 4.1), the involution $\tau_*$ induces an antiholomorphic involution of the space $H_{g,n}^{\mathbb{C}}$. The fixed points of this involution are the pseudoreal meromorphic functions.

Pseudoreal meromorphic functions can be studied following the same scheme as for real functions. In this way one can prove

THEOREM 5.8. *Let $(P, \tau, f)$ be a pseudoreal meromorphic function. Then $(P, \tau)$ is a real algebraic curve of topological type $(g, 0, 0)$ and the number of sheets (the degree) $n$ of $f$ is congruent to $g + 1$ (mod 2).*

THEOREM 5.9. *The space $H(g, n)$ of pseudoreal meromorphic functions of genus $g$ of degree $n$ is nonempty and connected for all $n \equiv g + 1$ (mod 2).*

# Bibliography

[1] N. L. Alling and N. Greenleaf, *Foundations of the theory of Klein surfaces.* Berlin–Heidelberg–New York. Springer-Verlag, 1971. Lecture Notes in Mathematics, vol. 219.

[2] C. Arf, *Untersuchungen über quadratische formen in Körpern der Charakteristik 2.* I, J. Reine Angew. Math. **183** (1941), 148–167.

[3] V. I. Arnold, *Topological classification of complex trigonometric polynomials and the combinatorics of graphs with an identical number of vertices and edges.* Funct. Anal. Appl. **30** (1996), no. 1, 1–17.

[4] M. F. Atiyah, *Riemann surfaces and spin structures.* Ann. Sci. École Norm. Sup. (4). **4** (1971), 47–62.

[5] H. F. Baker, *Abel's theorem and the allied theory including the theory of theta functions.* Cambridge, 1897.

[6] A. M. Baranov and A. S. Shvarts [Schwarz], *Multiloop contribution to string theory.* JETP Lett. **42** (1985), no. 8, 419–421.

[7] A. M. Baranov, Yu. I. Manin, I. V. Frolov, and A. S. Schwarz, *A superanalog of the Selberg trace formula and multiloop contributions for fermionic strings.* Comm. Math. Phys. **111** (1987), 373–392.

[8] L. Bers, *Quasiconformal mapping and Teichmüller's theorem.* in: Analytic Functions. Princeton Univ. Press, 1960. pp. 89–119. Princeton Math. Ser. vol. 24.

[9] A. I. Bobenko, *Uniformization and finite-gap integration,* preprint LOMI, P10-86, Leningrad, 1986.

[10] ———, *Schottky uniformization and finite-gap integration.* Soviet Math. Dokl. **36** (1988), no. 1, 38–42.

[11] W. Burnside, *On a class of automorphic functions.* Proc. London Math. Soc. **23** (1892), 49–88.

[12] E. Bujalance, A. F. Costa, S. Natanzon, and D. Singerman, *Involutions of compact Klein surfaces.* Mathematische Zeitschrift **211** (1992), no. 3, pp. 461–478.

[13] A. L. Carey and K. C. Hannabuss, *Infinite dimensional groups and Riemann surface field theories.* Comm. Math. Phys. **176** (1996), no. 2, 321–351.

[14] I. V. Cherednik, *On the conditions of reality in "finite-gap integration".* Dokl. Akad. Nauk SSSR **252** (1980), no. 5, 1104–1108.

[15] A. Clebsch, *Zur Theorie der Riemann Flächer.* Math. Ann. **6** (1873), 1–15.

[16] A. Comessatti, *Sulle variata abeliane reali.* Ann. Math. Pura Appl. (4). **2** (1925), 67–102; **3** (1926), 27–71.

[17] M. Dehn, *Über Abbildungen.* Math. Tidsskr. B. **1939**, 25–48.

[18] B. A. Dubrovin, *Theory of operators and real algebraic geometry.* in: Global analysis—studies and applications, III. Berlin: Springer-Verlag, 1988. pp. 42–59. Lecture Notes in Mathematics, vol. 1334.

[19] ———, *Geometry of 2D topological field theories.* in: Integrable Systems and Quantum Groups. Berlin: Springer-Verlag, 1996. pp. 120–348. Lecture Notes in Mathematics, vol. 1620.

[20]  _____, *Painlevé transcendents in two-dimensional topological field theory*. Preprint SISSA. 1998. 24/98/FM.

[21]  B. A. Dubrovin and S. M. Natanzon, *Real two-zone solutions of the sine-Gordon equation*. Funktsional. Anal. i Prilozhen. **16** (1982), no. 1, 27–43, 96.

[22]  _____, *Real theta-function solutions of the Kadomtsev-Petviashvili equation*. Math. USSR-Izv. **32** (1989), no. 2, 269–288.

[23]  B. A. Dubrovin, S. P. Novikov, and A. T. Fomenko, *Modern geometry—methods and applications*. Springer-Verlag, New York, 1992.

[24]  C. J. Earle, *On the moduli of closed Riemann surfaces with symmetries*. in: Advances in the theory of Riemann surfaces. Princeton Univ. Press, 1971. pp. 119–130. Ann. of Math. Stud. no. 66.

[25]  G. Falqui and C. Reina, *N = 2 super Riemann surfaces and algebraic geometry*. J. Math. Phys. **31** (1990), no. 4, 948–952.

[26]  J. Fay, *Theta-functions on Riemann surfaces*. Berlin: Springer-Verlag, 1973. Lecture Notes in Mathematics, vol. 352.

[27]  D. Friedan, *Notes on string theory and two-dimensional conformal field theory*. in: Proc. Workshop on unified string theories (Santa Barbara, Calif., 1985). Singapore: World Sci. Publishing, 1986.

[28]  F. Frike, F. Klein, *Vorlesungen über die Theorie der automorphen Funktionen*. B. 1, 2. Leipzig: Teubner, 1897, 1912. — New York: Johnson Reprint Corp., and Stuttgart: Teubner Verlagsgesellschaft, 1965.

[29]  D. Grepner, *Space-time supersymmetry in compactified string theory and superconformal models*. Nuclear Phys. **296** (1988), 757–779.

[30]  B. H. Gross and J. Harris, *Real algebraic curves*. Ann. Sci. École Norm. Sup. (4) **14** (1981), no. 2, 157–182.

[31]  A. Harnack, *Über die Vieltheiligkeit der ebenen algebraischen Kurven*. Math. Ann. **10** (1876), 189–199.

[32]  A. Hurwitz, *Über Riemannsche Flächen mit gegebenen Verzweigungspunkten*. Math. Ann. **39** (1891), 1–61.

[33]  _____, *Über die Fourierschen Konstanten integrierbarer Funktionen*. Math. Ann. **57** (1903), 425–446.

[34]  A. Jaffe, S. Klimek, and L. Lesniewski, *Representations of the Heisenberg algebra on a Riemann surface*. Comm. Math. Phys. **126** (1990), no. 2, 421–433.

[35]  D. Johnson, *Spin structures and quadratic forms on surfaces*. J. London Math. Soc. (2) **22** (1980), no 2, 365–373.

[36]  L. Keen, *Intrinsic moduli on Riemann surfaces*. Ann. of Math. (2) **84** (1966), no. 3, 404–420.

[37]  _____, *Canonical polygons for finitely generated Fuchsian groups*. Acta Math. **115** (1965), 1–16.

[38]  _____, *On Fricke moduli*. in: Advances in the Theory of Riemann Surfaces. Princeton Univ. Press, 1971, pp. 205–224. Ann. of Math. Studies. No. 66. *A correction to "On Fricke moduli"*, Proc. Amer. Math. Soc. **40** (1973), 60–62.

[39]  B. Kerekjarto, *Vorlesungen über Topologie. I. Flächentopologie*. Berlin: Springer-Verlag, 1923.

[40]  F. Klein, *Riemann Flächen*. Göttingen: Vorlesung. 1892, vol. 1, 2; Neuedruck, 1906.

[41]  M. Kontsevich and Yu. Manin, *Gromov–Witten classes, quantum cohomology and enumerative geometry*. Comm. Math. Phys. **164** (1994), no. 3, 525–562.

[42]  S. Kravetz, *On the geometry of Teichmüller spaces and the structure of their modular groups*. Ann. Acad. Sci. Fenn. Ser. A I. **278** (1959), 1–35.

[43]  I. M. Krichever and S. P. Novikov, *Virasoro–Gelfand–Fuks type algebras, Riemann surfaces, operator's theory of closed strings*. J. Geom. Phys. **5** (1988), no. 4, 631–661.

[44] A. M. Macbeath, *The classification of non-Euclidean plane crystallographic groups.* Canad. J. Math. **19** (1967), no. 6, 1192–1205.

[45] Yu. I. Manin, *Superalgebraic curves and quantum strings.* in: Proc. Steklov Inst. Math. **1991**, no. 4, 149– 162.

[46] Yu. Manin and S. Merkulov, *Semisimple Frobenius (super) manifolds and quantum cohomology of $P^r$.* Topol. Methods Nonlinear Anal. **9** (1997), no. 1, 107–161.

[47] A. D. Mednykh, *Determination of the number of nonequivalent coverings over a compact Riemann surface.* Dokl. Akad. Nauk SSSR **239** (1978), no. 2, 269–271.

[48] M. I. Monastyrsky and S. M. Natanzon, *The moduli space of superconformal instantons in sigma models.* Modern Phys. Lett. A. **6** (1991), no. 19, 1787–1796; ibid., no. 31, 2919.

[49] _____, *The moduli space of instantons in $N = 2$ supersymmetrical $\sigma$-models.* Classical Quantum Gravity **12** (1995), no. 9, 2149–2156; **13** (1996), 1277.

[50] _____, *The moduli space of instantons in $N = 2$ supersymmetrical $\sigma$-models.* in: Topics in statistical and theoretical physics. Providence, RI: Amer. Math. Soc., 1996, pp. 195–202. Amer. Math. Soc. Transl. Ser. 2, vol. 177.

[51] D. Mumford, *Theta characteristics of an algebraic curve.* Ann. Sci. École Norm. Sup. (4) **4** (1971), 181–192.

[52] S. M. Natanzon, *Invariant lines of Fuchsian groups.* Uspekhi Mat. Nauk **27** (1972), no. 4(166), 145– 160.

[53] _____, *Invariant lines of Fuchsian groups and moduli of real algebraic curves.* PhD Thesis, Moscow, TsEMI, 1974.

[54] _____, *Moduli of real algebraic curves.* Uspekhi Mat. Nauk **30** (1975), no. 1(181), 251–252.

[55] _____, *Moduli spaces of real algebraic curves.* in: Proc. Moscow Math. Soc. **37** (1978), 219–253.

[56] _____, *Finite groups of homeomorphisms of surfaces and real forms of complex algebraic curves.* PhD Thesis, Yaroslavl, 1980.

[57] _____, *Geometric description of the action of hyperelliptic involutions on the homology group of the surface.* in: Constructive algebraic geometry, no. 194, Yaroslavl State Pedagogical Institute, Yaroslavl, 1981, pp. 89-96.

[58] _____, *On the number and topological types of real hyperelliptic curves isomorphic over $\mathbb{C}$.* in: Constructive algebraic geometry, no. 200, Yaroslavl State Pedagogical Institute, Yaroslavl, 1982, pp. 82–93.

[59] Natanzon S. M., *Spaces of real meromorphic functions on real algebraic curves.* Dokl. Akad. Nauk SSSR **279** (1984), no. 4, 803–805.

[60] _____, *Uniformization of spaces of meromorphic functions.* Dokl. Akad. Nauk SSSR **287** (1986), no. 5, 1058–1061.

[61] _____, *Real meromorphic functions on real algebraic curves.* Soviet Math. Dokl. **36** (1988), no. 3, 425-427.

[62] _____, *The Fricke space of super-Fuchsian groups.* Funktsional. Anal. i Prilozhen. **21** (1987), no. 2, 80–81.

[63] _____, *Topology of two-dimensional coverings, and meromorphic functions on real and complex algebraic curves. I.* In: Trudy Sem. Vektor. Tenzor. Anal. No. 23 (1988), 79–103.

[64] _____, *The moduli space of Riemann supersurfaces.* Math. Notes **45** (1989), no. 3-4, 341–345.

[65] _____, *Prymians of real curves and their applications to the effectivization of Schrödinger operators.* Funct. Anal. Appl. **23** (1989), no. 1, 33–45.

[66] _____, *Klein surfaces.* Russian Math. Surveys **45** (1990), no. 6, 53–108.

[67] _____, *Klein supersurfaces.* Math. Notes **48** (1990), no. 1-2, 766–772 (1991).

[68] _____, *Topology of two- dimensional coverings, and meromorphic functions on real and complex algebraic curves. II.* In: Trudy Sem. Vektor. Tenzor. Anal. No. 24 (1991), 104–132.

[69] _____, *Supercoverings, SNEC-groups and inner groups of Riemann and Klein supersurfaces.* Russian Math. Surveys **45** (1990), no. 2, 225–226.

[70] _____, *Discrete subgroups of* GL(2, C) *and spinor bundles on Riemann and Klein surfaces.* Funct. Anal. Appl. **25** (1991), no. 4, 293–294 (1992).

[71] _____, *Differential equations for Prym theta functions. A criterion for two-dimensional finite-gap potential Schrodinger operators to be real.* Funct. Anal. Appl. **26** (1992), no. 1, 13–20.

[72] _____, *Topological invariants and moduli of hyperbolic* $N = 2$ *Riemannian supersurfaces.* Acad. Sci. Sb. Math. **79** (1994), no. 1, 15–31.

[73] _____, *Moduli spaces of Riemann and Klein supersurfaces.* in: Developments in Mathematics: The Moscow School. London: Chapman & Hall, 1993, pp. 100–130.

[74] _____, *Moduli spaces of Riemann* $N = 1$ *and* $N = 2$ *supersurfaces.* J. Geom. Phys. **12** (1993), no. 1, 35–54.

[75] _____, *Topology of 2-dimensional coverings and meromorphic functions on real and complex algebraic curves.* Selecta Math. Soviet. **12** (1993), no. 3, 251–291.

[76] _____, *Classification of pairs of Arf functions on orientable and nonorientable surfaces.* Funct. Anal. Appl. **28** (1994), no. 3, 178–186.

[77] _____, *On quadratic forms over the field* $Z_2$. Russian Math. Surveys **50** (1995), no. 5, 1090–1091.

[78] _____, *Real nonsingular finite zone solutions of soliton equations.* in: Topics in topology and mathematical physics. Providence, RI: Amer. Math. Soc., 1995, pp. 153–183. Amer. Math. Soc. Transl. Ser. 2, vol. 170.

[79] _____, *Trigonometric tensors on algebraic curves of arbitrary genus. An analogue of the Sturm–Hurwitz theorem.* Russian Math. Surveys **50** (1995), no. 6, 1286–1287.

[80] _____, *Spinors and differentials of real algebraic curves.* in: Topology of real algerbraic varieties and related topics. Providence, RI: Amer. Math. Soc., 1996, pp. 179–186. Amer. Math. Soc. Transl. Ser. 2, vol. 173.

[81] _____, *Differential equations for Riemann and Prym theta-functions.* J. Math. Sci. **82** (1996), no. 6, 3821–3823.

[82] _____, *Moduli spaces of real algebraic supercurves with* $N = 2$. Funct. Anal. Appl. **30** (1996), no. 4, 237–245 (1997).

[83] _____, *Spaces of meromorphic functions on Riemann surfaces.* in: Topics in singularity theory. Providence, RI: Amer. Math. Soc., 1997, pp. 175–180. Amer. Math. Soc. Transl. Ser. 2, vol. 180.

[84] _____, *The topological structure of the space of holomorphic morphisms of Riemann surfaces.* Russian Math. Surveys **53** (1998), no. 2, 398–400.

[85] _____, *Moduli of Riemann surfaces of a Hurwitz-type space and their superanalogues.* Russian Math. Surveys **54** (1999), no. 1, 61–117.

[86] _____, *Moduli of real algebraic curves and their superanalogues. Spinors and Jacobians of real curves.* Russian Math. Surveys **54** (1999), no. 6, 1091–1147.

[87] A. M. Polyakov, *Quantum geometry of bosonic string.* Phys. Lett. **103** (1981), 207.

[88] A. A. Rosly, A. S. Schwarz, and A. A. Voronov, *Geometry of superconformal manifolds.* Comm. Math. Phys. **119** (1988), no. 4, 129–152.

[89] Y. Ruan and G. Tian, *A mathematical theory of quantum cohomology.* J. Differential Geom. **42** (1995), no. 2, 259–367.

[90] M. Seppälä, *Teichmüller spaces of Klein surfaces.* Ann. Acad. Sci. Fenn. Ser. A I Math. Dissertationes. 1978, no. 15.

[91] L. Vajsburd and A. Radul, *Non-orientable strings.* Comm. Math. Phys. **135** (1991), 413–420.

[92] V. Vinnikov, *Self-adjoint determinantal representations of real plane curves*. Math. Ann. **296** (1993), no. 3, 453–479.

[93] G. Weichold, *Über symmetrische Riemannsche Flächen und die Periodizitäts moduln der zugehörigen Abelschen Normalintegrale erster Gatt*. Zeitschrift für Math. und Phys. **28** (1883), 321–351.

[94] S. Wolpert, *The length spectra as moduli for compact Riemann surface*. Ann. of Math. (2). **109** (1979), no. 2, pp. 323–351.

[95] H. Zieschang, E. Vogt, and H. D. Coldewey, *Surfaces and planar discontinuous groups*. Berlin: Springer-Verlag, 1980. Lecture Notes in Mathematics, vol. 835.

[96] S. Zdravkovska, *The topological classification of polynomial mappings*. Uspekhi Mat. Nauk **25** (1970), no. 4(154), 179–180.

[97] A. Zhivkov, *Finite-gap matrix potential with one and two involutions*. Bull. Sci. Math. **118** (1994), no. 5, pp. 403–440.

# Index